BUDDHISM

BUDDHISM

A CONCISE INTRODUCTION

~~~

HUSTON SMITH
AND PHILIP NOVAK

HarperSanFrancisco
A Division of HarperCollinsPublishers

To all followers of the Dharma, and to others who are interested in exploring its potentials to improve individual lives and the course of history, the authors respectfully dedicate this book.

HarperCollins Web site: http://www.harpercollins.com
HarperCollins®, ☲®, and HarperSanFrancisco™ are
trademarks of HarperCollins Publishers, Inc.

FIRST HARPERCOLLINS PAPERBACK EDITION PUBLISHED IN 2004
Designed by Joseph Rutt

Library of Congress Cataloging-in-Publication Data
has been ordered.
ISBN 0–06–073067–6 (pbk.)
05 06 07 08 RRD(H) 10 9 8 7 6 5 4 3 2

CONTENTS

FOREWORD

This book reconceives the chapter on Buddhism from Huston Smith's *The World's Religions* and takes advantage of the additional space a book provides to go deeper into Buddhism's basics. Most important here, Theravada Buddhism (which was overshadowed by Mahayana when *The World's Religions* was written) is brought closer to getting its due. Then, on these foundations the book erects a second story, so to speak. The second half of the book, entirely new, tells the story of Buddhism's migration to the West, particularly to America.

In this happily co-authored book, the authors have worked over every page together with Smith taking the lead in its first half and Novak in its second. Then, for reasons that will be noted in due course, the lead swings back to Smith who wrote the Afterword on Pure Land Buddhism.

The partnership that went into the book proved to be a fortunate one in many ways. Apart from the fact that the talents of each author complement those of the other, Novak wrote his doctoral dissertation under Smith at Syracuse University, and geographical proximity—Novak teaches at Dominican University in San Rafael, a half-hour's drive across San Francisco Bay from Smith in Berkeley—has allowed their friendship to age like old wine. One proof of its vintage is that they felt comfortable in raising their voices at each other when disputes arose as they invariably must in joint authorship. In every case, however, the differences were resolved in ways that both parties felt led to a better book.

Yet another way the authors complement each other is that between them their Buddhist practices cover both sides of Buddhism. Novak has been a lifelong practitioner of Theravada *vipassana*, while Smith was for fifteen years a disciple of Goto Zuigan Roshi in Mahayana Zen.

The authors wish to thank the book's editor, John Loudon, for conceiving and commissioning this book. The inducement it provided for them to clear eight months to wash their minds and spirits once again through the treasures of this great tradition came as a great refreshment, staking out as it were an oasis in their busy lives. Long, long ago the Buddha embarked on a search for a way to live life fully and vibrantly while facing unflinchingly the inexorable axioms of aging, sickness, and death. By the time of his death he had found such a way, and in the 2,500 years since, it has transformed the lives of the millions who have followed him.

Two other notes need to be added. The first concerns terminology. Buddhist vocabulary has come down to us in two ancient Indian languages, Pali and Sanskrit, and Sanskrit terms like *karma, nirvana,* and *dharma* are more familiar in the West than their Pali versions, *kamma, nibbana,* and *dhamma.* One might conclude that exclusive use of Sanskrit terms in a book like this would be the obvious way to proceed. But the matter is trickier. Sometimes the reverse is true and Pali terms like *anicca* (impermanence) and *anatta* (no-self) are better known than the Sanskrit *anitya* and *anatman*. Accordingly, our general rule has been to honor familiar usage rather than attempt to maintain consistency with one language. Exceptions to this rule occur only in Chapters 8 and 18, both on Theravada Buddhism, where out of deference to that tradition's close connection to Pali we use only Pali terms. Second, with the exception of terms like karma and nirvana, which have become part of the West's vocabulary, we italicize foreign terms the first time we use them but not thereafter.

It remains for us to acknowledge the invaluable aid we have received from others. A certain writer has said that everybody except myself has been my mentor and we resonate with that assertion. However, there are certain individuals who have been special sources of help and encouragement in this project. We would like to thank Dhananjay Chavan, John Kling, Donald Rothberg, Harry and Vivian Snyder, and Roger Walsh for their valuable suggestions upon reading parts or all of the text during various phases of its completion. Great thanks is also due to our production editor, Chris Hafner, for superb supervision. Tetsuo Unno's help with the book's Afterword will be acknowledged there. Of course, any defects that remain in the text are solely the author's responsibility. Finally, Novak would like to thank Dominican University of California for the sabbatical leave that freed him to work on this project.

Huston Smith and Philip Novak
November 2002

ACKNOWLEDGMENTS

Acknowledgment is made to the following for permission to reprint copyrighted material:

From *Buddhist Texts Through the Ages,* © Muriel Conze, 1995. Reproduced with permission of Oneworld Publications.

From *Buddhist Scriptures,* translated by Edward Conze (Penguin Classics, 1959), © Edward Conze, 1959. Reproduced with permission of Penguin Books, Ltd.

From *Buddhism: Its Essence and Development* by Edward Conze (Birmingham, England: Windhorse Publications, 2001), © Janet Kavanaugh, with permission from Windhorse Publications.

From *The Awakening of the West: The Encounter of Buddhism and Western Culture* (1994) by Stephen Batchelor with permission of Parallax Press, Berkeley, California, www.parallax.org

From *Like a Dream, Like a Fantasy—the Zen Writings of Nyogen Sensaki.* Used with permission from The Zen Studies Society, 223 East 67th St., New York, NY 10021. Eido T. Shimano, Roshi, Abbot. www.zenstudies.org

From *The New Buddhism* by James William Coleman, © 2001 by James William Coleman. Used with permission of Oxford University Press, Inc.

From Nyanaponika Thera, trans., *Anguttara Nikaya: An Anthology,* part II in *The Wheel,* Nos. 208–11, 56 ff; "The Practice of Metta" in the Sutta Nipata 145–51, translated by Nanamoli Thera in *The Wheel,* no. 7, p. 19; *Majjhima Nikaya,* Sutta no. 7, "The Simile of the Cloth," translated by Nyanaponika Thera; *Digha Nikaya* 16 *(Mahaparinibbana Sutta),* Part 3, verse 61; Soma Thera, trans., *The Way of Mindfulness,* 6th revised ed., a translation of the *Sattipatthana Sutta* of the *Majjhima Nikaya,* p. 11 of online version, http://www.accesstoinsight.org. Courtesy—Buddhist Publication Society, Inc., Sri Lanka.

From *Peace is Every Step* by Thich Naht Hanh (New York: Bantam, 1991) with permission from Bantam Books, a division of Random House, Inc.

From *The Path of Light* by L. D. Barnett with permission of John Murray (Publishers) Ltd.

From *Inquiring Mind,* vol. 18, no. 1, Fall 2001, p. 38. Reprinted with permission of S. N. Goenka and Inquiring Mind, P. O. Box 9999, Berkeley, CA 94709, www.inquiringmind.com

From *Srimad Bhagavata: The Holy Book of God,* Skanda 9, Chapter 21, verse 12, with permission of Sri Ramakrishna Math, Mylapore, Chennai, India.

From *How the Swans Came to the Lake* by Rick Fields, © 1981, 1986, 1992 by Rick Fields. Reprinted by arrangement with Shambhala Publications, Inc., Boston, www.shambhala.com.

Words of Shunryu Suzuki, Roshi, from *Wind Bell,* Vol. 5, no. 3, Summer, 1968, with permission of the San Francisco Zen Center.

From *The Dhammapada,* translated by Eknath Easwaran, founder of the Blue Mountain Center of Meditation, © 1985. Reprinted with permission of Nilgiri Press, www.nilgiri.org.

From "Digital Dharma," by Erik Davis, *Wired Online,* August, 1994, © 1994 Conde Nast Publications. All rights reserved. Originally published in *Wired.* Reprinted with permission.

From *The Buddhist Tradition* by William Theodore de Bary, © 1969 by William Theodore de Bary. Reprinted with permission of Random House, Inc.

From *The History of Zen Buddhism* by Heinrich Dumoulin (Boston: Beacon Press, 1963; paperback reprint). Original rights: Random House. Reprinted with permission of Random House, Inc.

From *Brother to Dragons* by Robert Penn Warren, © 1979 by Robert Penn Warren. Reprinted by permission of William Morris Agency, Inc., on behalf of the author.

The authors have made every effort to trace the copyright holders of every extract of more than forty words in this book. If they have inadvertently overlooked any, they will be pleased to make the necessary arrangements at the first opportunity.

PART I

~~~

THE WHEEL OF
THE DHARMA

~:~:~:

THE MAN WHO WOKE UP

Buddhism begins with a man. In his later years, when India was afire with his message and kings themselves were bowing before him, people came to him even as they were to come to Jesus asking what he was.[1] How many people have provoked this question—not "Who are you?" with respect to name, origin, or ancestry, but *"What* are you? What order of being do you belong to? What species do you represent?" Not Caesar, certainly. Not Napoleon, or even Socrates. Only two: Jesus and Buddha. When the people carried their puzzlement to the Buddha himself, the answer he gave provided an identity for his entire message:

> *"Are you a god?" they asked.*
> "No."
> *"An angel?"*
> "No."
> *"A saint?"*
> "No."

"Then what are you?"
Buddha answered, "I am awake."

His answer became his title, for this is what "Buddha" means. The Sanskrit root *budh* denotes both "to wake up" and "to know." Buddha, then, means the "Enlightened One," or the "Awakened One." While the rest of the world was wrapped in the womb of sleep, dreaming a dream known as the waking state of human life, one of their number roused himself. Buddhism begins with a man who shook off the daze, the doze, the dreamlike vagaries of ordinary awareness. It begins with a man who woke up.

His life has become encased in loving legend. We are told that the worlds were flooded with light at his birth. The blind so longed to see his glory that they received their sight; the deaf and mute conversed in ecstasy of the things that were to come. Crooked became straight; the lame walked. Prisoners were freed from their chains, and the fires of hell were quenched. Even the cries of the beasts were hushed as peace encircled the earth. Only Mara, the Evil One, did not rejoice.

The historical facts of his life are roughly these: He was born around 563 B.C.E. in what is now Nepal, near the Indian border. His full name was Siddhartha Gautama of the Sakyas. Siddhartha was his given name, Gautama his surname, and Sakya the name of the clan to which his family belonged. His father was a king, but as there were then many kingdoms in the subcontinent of India, it would be more accurate to think of him as a feudal lord. By the standards of the day Siddhartha's upbringing was luxurious. "I was delicate, O monks, excessively delicate. I wore garments of silk and my attendants held a white umbrella over me. My unguents were always from Banaras." He appears to have been exceptionally handsome, for there are numerous references to "the perfection of his visible body." At sixteen he married a neighboring princess, Yasodhara, who bore a son whom they called Rahula.

He was, in short, a man who seemed to have everything: family, "the venerable Gautama is well born on both sides, of pure descent"; fine appearance, "handsome, inspiring trust, gifted with great beauty of complexion, fair in color, fine in presence, stately to behold"; wealth, "he had elephants and silver ornaments for his elephants." He had a model wife, "majestic as a queen of heaven, constant ever, cheerful night and day, full of dignity and exceeding grace," who bore him a beautiful son. In addition, as heir to his father's throne, he was destined for fame and power.

Despite all this there settled over him in his twenties a discontent that was to lead to a complete break with his worldly estate. The source of his discontent is impounded in the legend of the Four Passing Sights, one of the most celebrated calls to adventure in all world literature. When Siddhartha was born, so this story runs, his father summoned fortune-tellers to find out what the future held for his heir. All agreed that this was no usual child. His career, however, was crossed with one ambiguity. If he remained within the world, he would unify India and become its greatest conqueror, a *Chakravartin* ("Wheel-Turner"),[2] or Universal King. If, on the other hand, he forsook the world, he would become not a world conqueror, but a world redeemer. Faced with this option, his father determined to steer his son toward the former destiny. No effort was spared to keep the prince attached to the world. Three palaces and forty thousand dancing girls were placed at his disposal; strict orders were given that no ugliness intrude upon the courtly pleasures. Specifically, the prince was to be shielded from contact with sickness, decrepitude, and death; even when he went riding, runners were to clear the roads of these sights.

One day, however, an old man was overlooked, or (as some versions have it) miraculously incarnated by the gods to effect the needed lesson: a man decrepit, broken-toothed, gray-haired, crooked and bent of body, leaning on a staff, and trembling. That day Siddhartha learned the fact of old age. Though the king extended his guard, on a second ride Siddhartha encountered a body racked with disease, lying by the roadside; and on a third journey, a corpse. Finally, on a fourth occasion he saw a monk with shaven head, ochre robe, and bowl, and on that day he learned of the life of withdrawal from the world in search of freedom. It is a legend, this story, but like all legends it embodies an important truth, for the teachings of the Buddha show unmistakably that it was the body's inescapable involvement with disease, decrepitude, and death that made him despair of finding fulfillment on the physical plane. "Life is subject to age and death. Where is the realm of life in which there is neither age nor death?"

Once he had perceived the inevitability of bodily pain and passage, fleshly pleasures lost their charm. The singsong of the dancing girls, the lilt of lutes and cymbals, the sumptuous feasts and processions, the elaborate celebration of festivals only mocked his brooding mind. Flowers nodding in the sunshine and snows melting on the Himalayas cried louder of the evanescence of worldly things. He determined to quit the snare of distractions his palace had become and follow the call of a truth-seeker. One night in his twenty-ninth year he made the break, his Great Going Forth. Making his way in the post-midnight hours to where his wife and son were locked in sleep, he bade them both a silent good-bye, and then ordered the gatekeeper to bridle his great white horse.

The two mounted and rode off toward the forest. Reaching its edge at daybreak, Gautama changed clothes with the attendant, who returned with the horse to break the news. "Tell my father," said Gautama,

> *that there is no reason he should grieve. He will perhaps say it was too early for me to leave for the forest. But even if affection should prevent me from leaving my family just now of my own accord, in due course death would tear us apart, and in that we would have no say. Birds settle on a tree for a while, and then go their separate ways again. The meeting of all living beings must likewise inevitably end in their parting. This world passes away and disappoints the hopes of everlasting attachment. It is therefore unwise to have a sense of ownership for people who are united with us as in a dream—for a short while only and not in fact.*[3]

Then Gautama shaved his head and, "clothed in ragged raiment," plunged into the forest in search of enlightenment.

Six years followed, during which his full energies were concentrated toward this end. "How hard to live the life of the lonely forest dweller, to rejoice in solitude. Verily, the silent groves bear heavily upon the monk who has not yet won to fixity of mind!" The words bear poignant witness that his search was not easy. It appears to have moved through three phases, without record as to how long each lasted or how sharply the three were divided. His first act was to seek out two of the foremost Hindu masters of the day and pick their minds for the wisdom in their vast tradition. He learned a great deal—about *raja yoga*, the yoga of meditation, especially, but about Hindu philosophy as well;

so much in fact that Hindus came to claim him as their own, holding that his criticisms of the religion of his day were in the order of reforms and were less important than his agreements. In time, however, having mastered the deepest mystical states his teachers knew, he concluded that these yogis could teach him nothing more.

His next step was to join a band of ascetics and give their way an honest try. Was it his body that was holding him back? He would break its power and crush its interference. A man of enormous willpower, the Buddha-to-be outdid his associates in every austerity they proposed. He ate so little—six grains of rice a day during one of his fasts—that "when I thought I would touch the skin of my stomach I actually took hold of my spine." He would clench his teeth and press his tongue to his palate until "sweat flowed from my armpits." He would hold his breath until it felt "as if a strap were being twisted around my head."[4] In the end he grew so weak that he fell into a faint; and if a passing cowherdess had not stopped to feed him some warm rice gruel, he could easily have died.

This experience taught him the futility of asceticism. He had given this experiment all anyone could, and it had not succeeded—it had not brought enlightenment. But negative experiments carry their own lessons, and in this case asceticism's failure provided Gautama with the first constructive plank for his program: the principle of the Middle Way between the extremes of asceticism, on the one hand, and indulgence, on the other. It is the concept of the rationed life, in which the body is given what it needs to function optimally, but no more.

The experience also took his memory back to a day in his youth when, having wandered deep into the countryside, he sat down, quiet and alone, beneath an apple tree. The exertions of a farmer plowing a distant field bespoke the eternity of labor necessary to wrest sustenance from the earth. The sun's slow, ceaseless passage across the sky betokened the countless creatures in the air, on the earth, and under the ground that would soon perish. As he reflected steadily on life's impermanence, his mind opened onto a new state of lucid equanimity. It was now calm and pliable, and the clarity of its seeing was marred by neither elation nor sorrow. It was his first deep meditation—not an otherworldly trance, but a clear and steady seeing of the way things are. And more, it was accomplished in the normal conditions of life without needing to subject the body to starvation.

Having turned his back on mortification, Gautama now devoted the final phase of his quest to a combination of rigorous thought and deep concentration. One evening near Gaya in northeast India, south of the present city of Patna, he sat down under a peepul tree that has come to be known as the Bo Tree (short for *bodhi*, "enlightenment"). The place was later named the Immovable Spot, for tradition reports that the Buddha, sensing that a breakthrough was near, seated himself that epoch-making evening vowing not to arise until he was enlightened.

The records offer as the first event of the night a temptation scene reminiscent of Jesus' on the eve of his ministry. The Evil One, realizing that his antagonist's success was imminent, rushed to the spot to disrupt his concentrations. He attacked first in the form of Kama, the God of Desire, parading three voluptuous women with their tempting retinues. When the Buddha-to-be remained unmoved, the Tempter switched his guise to that of Mara, the Lord of Death. His powerful hosts assailed the aspirant with hurricanes, torrential rains, and showers of flaming rocks, but Gautama had so emptied himself of his finite self that the weapons found no target to strike and turned into flower petals as they entered his field of concentration. When, in final desperation, Mara challenged his right to do what he was doing, Gautama touched the earth with his right fingertip, whereupon the earth responded, "I bear you witness," with a hundred, a thousand, and a hundred thousand thunderous roars. Mara's army fled in rout, and the gods of heaven descended in rapture to tend the victor with garlands and perfumes.

Thereafter, while the Bo Tree rained red blossoms that full-mooned May night, Gautama's meditation steadily deepened. During the first watch[5] of the night, Gautama saw, one by one, his many thousands of previous lifetimes. During the second watch, his vision widened. It surveyed the death and rebirth of the whole universe of living beings and noted the ubiquitous sway of the law of *karma*—that good actions lead to happy rebirths, bad actions to miserable ones. During the third watch, Gautama saw what made the whole thing go: the universal law of causal interdependence. He called it *dependent arising*, and later identified it as the very heart of his message.[6] Thus armed, he made quick work of the last shreds of ignorant clinging that bound him to the wheel of birth and death.

As the morning star glittered in the transparent sky of the east, his mind pierced at last the bubble of the universe and shattered it to naught, only, wonder of wonders, to find it miraculously restored with the effulgence of true being. The Great Awakening had occurred. Freedom was his. His being was transformed, and he emerged the Buddha. From the center of his joy came a song of spiritual victory:

> *Through many a birth I wandered in this world,*
> *Seeking in vain the builder of this house.*
> *Unfulfilling it is to be born again and again!*

O housemaker! Now I have seen you!
You shall build no more houses for me!
Your beams are broken,
Your ridgepole is shattered.
My mind is free from all past conditionings,
And craves the future no longer.[7]

The event had cosmic import. All created things filled the morning air with their rejoicings, and the earth quaked six ways with wonder. Ten thousand galaxies shuddered in awe as lotuses bloomed on every tree, turning the entire universe into "a bouquet of flowers set whirling through the air."[8] The bliss of this vast experience kept the Buddha rooted to the spot for seven entire days. On the eighth he tried to rise, but another wave of bliss broke over him. For a total of forty-nine days he was lost in rapture, after which his "glorious glance" opened onto the world.

Mara was waiting for him again with one last temptation. He appealed this time to what had always been Gautama's strong point, his reason. Mara did not argue the burden of reentering the world, with its banalities and obsessions. He posed a deeper challenge. Who could be expected to understand truth as profound as that which he, the Buddha, had laid hold of? How could speech-defying revelation be compressed into words, or visions that shatter definitions be caged in language? In short, how to show what can only be found; teach what can only be learned? Why bother to play the idiot before an uncomprehending audience? Why not wash one's hands of the whole hot world, be done with the body and slip at once into *nirvana,* the blissful state of liberation from the cycle of death and rebirth. The argument was so persuasive that it almost carried the day. At length, however, the Buddha answered, "There will be some whose eyes are only slightly dimmed by dust, and they will understand." With this, Mara was banished from his life forever.

Nearly half a century followed, during which the Buddha trudged the dusty paths of India preaching his ego-shattering, life-redeeming message until his hair was white, step infirm, and body nothing but a burst drum. He founded an order of monks and nuns—now the oldest historical institution on our planet—challenged the deadness of a society forged by the dominating *brahmins* (the Hindu priestly caste), and accepted in return the resentment, queries, and bewilderment his stance provoked. His daily routine was staggering. In addition to training monks and overseeing the affairs of his order, he maintained an interminable schedule of public preaching and private counseling, advising the perplexed, encouraging the faithful, and comforting the distressed. "To him people come right across the country from distant lands to ask questions, and he bids all welcome." Underlying his response to these pressures and enabling him to stand up under them was the pattern of withdrawal and return that is basic to all creativity. The Buddha withdrew for six years; then returned for forty-five. But each year was likewise divided: nine months in the world, followed by a three-month retreat with his monks during the rainy season. His daily cycle, too, was patterned to this mold. His public hours were long, but three times a day he withdrew, to return his attention (through meditation) to its sacred source.

After an arduous ministry of forty-five years, at the age of eighty and around the year 483 B.C.E., the Buddha died from dysentery after eating a meal of dried boar's flesh in the home of Cunda the smith. Even on his deathbed his mind moved toward others. In the midst of his pain, it occurred to him that Cunda might feel responsible for his death. His last request, therefore, was that Cunda be informed that of all the meals he had eaten during his long life, only two stood out as having blessed him exceptionally. One was the meal whose strength had enabled him to reach enlightenment under the Bo Tree, and the other the one that was opening to him the final gates to nirvana. The many who approached his deathbed unable to contain their tears he chastised gently: "In the hour of joy, it is not proper to grieve." These are but two of the scenes that *The Book of the Great Decease* has preserved. Together they present a picture of a man who passed into the state in which "ideas and consciousness cease to be" without the slightest resistance. Two sentences from his valedictory have echoed through the ages: "All compounded things decay. Work out your own salvation with diligence."

~-~-~-

THE SILENT SAGE

To understand Buddhism it is of utmost importance to gain some sense of the impact of Buddha's life on those who came within its orbit.

It is impossible to read the accounts of that life without emerging with the impression that one has been in touch with one of the greatest personalities of all time. The obvious veneration felt by almost all who knew him is contagious, and the reader is soon caught up with his disciples in the sense of being in the presence of something close to wisdom incarnate.

Perhaps the most striking thing about him was his combination of a cool head and a warm heart, a blend that shielded him from sentimentality, on the one hand, and indifference, on the other. He was undoubtedly one of the greatest rationalists of all times, resembling in this respect no one as much as Socrates. Every problem that came his way was automatically subjected to cool, dispassionate analysis. First, it would be dissected into its component parts, after which these would be reassembled in logical, architectonic order with their meaning and import laid bare. He was a master of dialogue and dialectic, and calmly confident. "That in disputation with anyone whomsoever I could be thrown into confusion or embarrassment—there is no possibility of such a thing."

The remarkable fact, however, was the way this objective, critical component of his character was balanced by a Franciscan tenderness so strong as to have caused his message to be subtitled "a religion of infinite compassion." Whether he actually risked his life to free a goat that was snagged on a precipitous mountainside may be historically uncertain, but the act would certainly have been in character, for his life was one continuous gift to the famished crowds. Indeed, his self-giving so impressed his biographers that they could explain it only in terms of a momentum that had acquired its trajectory in the animal stages of his incarnations. The *Jataka Tales*[1] have him sacrificing himself for his herd when he was a stag, and hurling himself as a hare into a fire to feed a starving brahmin. Dismiss these *post facto* accounts as legends if we must; there is no question but that in his life as the Buddha the springs of tenderness gushed abundant. Wanting to draw the arrows of sorrow from everyone he met, he gave to each his sympathy, his enlightenment, and the strange power of soul, which, even when he barely spoke a word, gripped the hearts of his visitors and left them transformed.

Such was the experience of the young woman Kisa Gotami, who had found in her newborn son the full measure of life's joy and fulfillment, until the infant suddenly died. Deranged by grief, she continued to carry the child on her hip as she went from house to house asking for medicine to cure him. Someone took pity and sent her to the Buddha. "O Exalted One," she said, "give me medicine for my son!" The

Buddha replied that she had done well to come to him for medicine. He told her to return to town to fetch a few mustard seeds from each household where no one had died and to bring the collection back to him. Relieved that a magical ritual for her son's resurrection was under way, she set out eagerly. An exhaustive canvass, however, yielded not a single grain. At every door in town the reply was the same: "O, Gotami, many have died here!" Finally Kisa Gotami realized the nature of the medicine Buddha had dispensed. Her insane grief now replaced by gratitude for the Buddha's compassionate wisdom, she took her son to the cremation grounds.

In social encounters, the Buddha's royal upbringing stood him in good stead. "Fine in presence," he moved among kings and potentates with ease, for he had been one of them. Yet his poise and sophistication seem not to have distanced him from simple villagers. Surface distinctions of class and caste meant so little to him that he often appears not to have even noticed them. Regardless of how far individuals had fallen or been rejected by society, they received from the Buddha a respect that stemmed from the simple fact that they were fellow human beings. Thus many an outcaste and derelict, encountering for the first time the experience of being understood and accepted, found self-respect emerging and gained status in the community. "The venerable Gautama bids everyone welcome, is congenial, conciliatory, not supercilious, accessible to all."[2]

There was indeed an amazing simplicity about this man before whom kings bowed. Even when his reputation was at its highest, he would be seen, begging bowl in hand, walking through streets and alleys with the patience of one who knows the illusion of time. Like those of the vine and olive, two of the most symbolic plants that grow from the meagerest of soils, his physical needs were minimal. Once at Alavi during the frosts of winter he was found resting in meditation on a few leaves gathered on a cattle path. "Rough is the ground trodden by the hoofs of cattle; thin is the couch; light the monk's yellow robe; sharp the cutting wind of winter," he admitted. "Yet I live happily with sublime uniformity."

It is perhaps inaccurate to speak of Buddha as a modest man. John Hay, who was President Lincoln's secretary, said it was absurd to call Lincoln modest, adding that "no great human being is modest." Certainly, the Buddha felt that he had risen to a plane of understanding that was far above that of anyone else in his time. In this respect he simply accepted his superiority and lived in the self-confidence this acceptance bequeathed. But this is different from vanity or humorless conceit. At the final assembly of one of his *sangha*'s (community's) annual retreats, the Exalted One looked around over the silent company and said, "Well, ye disciples, I summon you to say whether you have any fault to find with me, whether in word or in deed." And when a favorite pupil exclaimed, "Such faith have I, Lord, that methinks there never was nor will be nor is now any other greater or wiser than the Blessed One," the Buddha admonished:

> "Of course, Sariputta, you have known all the
>   Buddhas of the past."
> "No, Lord."
> "Well then, you know those of the future?"
> "No, Lord."
> "Then at least you know me and have penetrated
> my mind thoroughly?"
> "Not even that, Lord."
> "Then why, Sariputta, are your words so grand and
> bold?"

Notwithstanding his own objectivity toward himself, there was constant pressure during his lifetime to turn him into a god. He rebuffed all attempts categorically, insisting that he was human in every respect. He made no attempt to conceal his temptations and weaknesses—how difficult it had been to attain enlightenment, how narrow the margin by which he had won through, how fallible he still remained. He confessed that if there had been another drive as powerful as sex, he would never have made the grade. He admitted that the months when he was first alone in the forest had brought him to the brink of mortal terror. "As I tarried there, a deer came by, a bird caused a twig to fall, and the wind set all the leaves whispering; and I thought, 'Now it is coming—that fear and terror.'" As Paul Dahlke remarks in his *Buddhist Essays,* "One who thus speaks need not allure with hopes of heavenly joy. One who speaks like this of himself attracts by that power with which the Truth attracts all who enter her domain."

Buddha's leadership was evidenced not only by the size to which his order grew, but equally by the perfection of its discipline. A king visiting one of their assemblies, which was prolonged into a full-moon night, burst out at last, "You are playing me no tricks? How can it be that there should be no sound at all, not a sneeze, nor a cough, in so large an Assembly, among 1,250 of the Brethren?" Watching the Assembly, seated as silent as a clear lake, he added, "Would that my son might have such calm."

Like other spiritual geniuses—one thinks of Jesus spotting Zacchaeus, a nobody who had perched himself in a tree for a glimpse of Jesus amid the pressing throngs and in whom Jesus sensed such great sincerity that he invited him into his inner circle—the Buddha was gifted with preternatural insight into character. Able to size up, almost at sight, the people who approached him, he seemed never to be taken in by fraud and front, but would move at once to what was authentic and genuine. One of the most beautiful instances of this was his encounter with Sunita the flower scavenger, a man so low on the social scale that the only employment he could find was picking over discarded bouquets to find an occasional blossom that might be bartered to still his hunger. When the Buddha arrived one day at the place where he was sorting through refuse, Sunita's heart was filled with awe and joy. Finding no place to hide—for he was an outcaste—he stood as if stuck to the wall, saluting with clasped hands. The Buddha "marked the conditions of arhatship [sainthood] in the heart of Sunita, shining like a lamp within a jar," and drew near, saying, "Sunita, what to you is this wretched mode of living? Can you endure to leave the world?" Sunita, "experiencing the rapture of one who has been sprinkled with ambrosia, said, 'If such as I may become a monk of yours, may the Exalted One suffer me to come forth!'" He became a renowned member of the order.[3]

The Buddha's entire life was saturated with the conviction that he had a cosmic mission to perform. Immediately after his enlightenment he saw in his mind's eye "souls whose eyes were scarcely dimmed by dust and souls whose eyes were sorely dimmed by dust"[4]—the whole world of humanity, milling, lost, desperately in need of help and guidance. He had no alternative but to agree with his followers that he had been "born into the world for the good of the many, for the happiness of the many, for the advantage, the good, the happiness of gods and men, out of compassion for the world."[5] His acceptance of this mission without regard for personal cost won India's heart as well as her mind. "The monk Gautama has gone forth into the religious life, giving up the great clan of his relatives, giving up much money and gold, treasure both buried and above ground. Truly while he was still a young man without gray hair on his head, in the beauty of his early manhood he went forth from the household life into the homeless state."[6]

Encomiums to the Buddha crowd the texts, one reason undoubtedly being that no description ever satisfied his disciples completely. After words had done their best, there remained in their master the essence of mystery—unplumbed depths their language could not express because thought could not fathom them. What they could understand they revered and loved, but there was more than they could hope to exhaust. To the end he remained half light, half shadow, defying complete intelligibility. So they called him Sakyamuni, "silent sage [muni] of the Sakya clan," symbol of something beyond what could be said and thought. And they called him Tathagata, the "Thus-gone,"[7] the "Truth-winner," the "Perfectly Enlightened One," for "he alone thoroughly knows and sees, face to face, this universe." "Deep is the Tathagata, unmeasurable, difficult to understand, even like the ocean."[8]

~·~·~

THE REBEL SAINT

In moving from Buddha the man to Buddhism the religion, it is imperative that the latter be seen against the background of the Hinduism out of which it grew. Unlike Hinduism, which emerged by slow, largely imperceptible spiritual accretion, the religion of the Buddha appeared overnight, fully formed. In large measure it was a religion of reaction against Hindu perversions—an Indian protestantism not only in the original meaning of that word, which emphasized witnessing for (Lat., *testis pro*) something, but equally in its latter-day connotations, which emphasize protesting against something. Buddhism drew its lifeblood from Hinduism, but against its prevailing corruptions Buddhism recoiled like a whiplash and hit back—hard.

To understand the teachings of the Buddha, then, we shall need a minimal picture of the existing Hinduism that partly provoked it. And to lead into this, several observations about religion are in order.

Six aspects of religion surface so regularly as to suggest that their seeds are in the human makeup. One of these is *authority*. Leaving divine authority aside and approaching the matter in human terms only, the point begins with specialization. Religion is no less complicated than government or medicine. It stands to reason, therefore, that talent and sustained attention will lift some people above the average in matters of spirit; their advice will be sought and their counsels generally followed. In addition, religion's institutional, organized side calls for administrative bodies and individuals who occupy positions of authority, whose decisions carry weight.

A second normal feature of religion is *ritual*, which was actually religion's cradle, for anthropologists tell us that people danced out their religion before they thought it out. Religion arose out of celebration and its opposite, bereavement, both of which cry out for collective expression. When we are crushed by loss, or when we are exuberant, we want not only to be with people; we want to interact with them in ways that make the interactions more than the sum of their parts—this relieves our isolation. The move is not limited to the human species. In northern Thailand, as the rising sun first touches the treetops, families of gibbons sing half-tone descending scales in unison as, hand over hand, they swoop across the topmost branches.

Religion may begin in ritual, but explanations are soon called for, so *speculation* enters as a third religious feature. Whence do we come, whither do we go, why are we here? People want answers to these questions.

A fourth constant in religion is *tradition*. In human beings it is tradition rather than instinct that conserves what past generations have learned and bequeath to the present as templates for action.

A fifth typical feature of religion is *grace*, the belief—often difficult to sustain in the face of facts—that Reality is ultimately on our side. In the last resort the universe is friendly; we can feel at home in it. "Religion says that the best things are the more eternal things, the things in the universe that throw the last stone, so to speak, and say the final word."[1]

Finally, religion traffics in *mystery*. Being finite, the human mind cannot begin to fathom the Infinite, which it is drawn to.

Each of these six things—authority, ritual, speculation, tradition, grace, and mystery—contributes importantly to religion, but equally each can clog its works. In the Hinduism of the Buddha's day they had done so, all six of them. Authority, warranted at the start, had become hereditary and exploitative as brahmins took to hoarding their religious secrets and charging exorbitantly for their ministrations. Rituals became mechanical means for obtaining miraculous results. Speculation had lost its experiential base and devolved into meaningless hair-splitting. Tradition had turned into a dead weight, in one specific by insisting that Sanskrit—no longer understood by the masses—remain the language of religious discourse. God's grace was being misread in ways that undercut human responsibility, if indeed responsibility any longer had meaning where karma, likewise misread, was confused with fatalism. Finally, mystery was confused with mystery-mongering and mystification—perverse obsession with miracles, the occult, and the fantastic.

Onto this religious scene—corrupt, degenerate, and irrelevant, matted with superstition and burdened with worn-out rituals—came the Buddha, determined to clear the ground that truth might find new life. The consequence was surprising. What emerged was (at the start) a religion almost entirely devoid of each of the above mentioned ingredients, without which we would suppose that religion could not take root. This fact is so striking that it warrants documentation.

1. Buddha preached a religion devoid of authority. His attack on authority had two prongs. On the one hand, he wanted to break the monopolistic grip of the brahmins on religious teachings, and a good part of his reform consisted of no more than making generally accessible what had hitherto been the possession of a few. Contrasting his own openness with the guild secrecy of the brahmins, he pointed out that "there is no such thing as closed-fistedness in the Buddha." So important did he regard this difference that he returned to it on his deathbed to assure those about him, "I have not kept anything back."[2]

But if his first attack on authority was aimed at an institution—the brahmin caste—his second was directed toward individuals. In a time when the multitudes were passively relying on brahmins to tell them what to do, Buddha challenged each individual to do his or her own religious seeking and rational investigation. "Do not go upon what has been acquired by repeated hearing; nor upon tradition; nor upon rumour; nor upon what is in a scripture; nor upon the consideration, 'The monk is our teacher.'"[3] Rather, he said, test ideas and actions in your own laboratory of common sense: When you yourself know they lead to harm or ill, abandon them; when you yourself know they lead to benefit and happiness, adopt them. Self-reliance was key: "Be lamps unto yourselves. Those who, either now or after I am dead, shall rely upon themselves only and not look for assistance to anyone besides themselves, it is they who shall reach the topmost height."[4]

2. Buddha preached a religion devoid of ritual. Repeatedly, he ridiculed the rigmarole of brahmin rites as superstitious petitions to ineffectual gods. "To seek to win peace

through others, as priests and sacrificers, is the same as if a stone were thrown into deep water, and now people, praying and imploring and folding their hands, came and knelt down all around saying: 'Rise, O dear stone! Come to the surface, O dear stone!' But the stone remains at the bottom."[5]

Rituals were trappings—irrelevant to the hard, demanding job of ego-reduction. Indeed, they were worse than irrelevant; Buddha argued that "belief in the efficacy of rites and ceremonies" is one of the Ten Fetters[6] that bind the human spirit. Here, as apparently everywhere, the Buddha was consistent. Discounting Hinduism's forms, he resisted every temptation to institute new ones of his own, a fact that has led some writers to (mistakenly) characterize his teachings as a rational moralism rather than a religion.

3. Buddha preached a religion that skirted speculation. There is ample evidence that he could have been one of the world's great metaphysicians if he had put his mind to the task. Instead, he skirted "the thicket of theorizing." His silence on that front did not pass unnoticed. "Whether the world is eternal or not eternal, whether the world is finite or not, whether the soul is the same as the body or whether the soul is one thing and the body another, whether a Buddha exists after death or does not exist after death—these things," one of his disciples observed, "the Lord does not explain to me. And that he does not explain them to me does not please me, it does not suit me."[7] There were many it did not suit. Yet despite incessant needling, he maintained his "noble silence." His reason was simple. On questions of this sort, "greed for views tends not to edification."[8] His practical program was exacting, and he was not going to let his disciples be diverted from the hard road of practice into fields of fruitless speculation. His famous parable of the arrow thickly smeared with poison puts the point with precision:

*It is as if a man had been wounded by an arrow
thickly smeared with poison, and his friends and
kinsmen were to get a surgeon to heal him, and he
were to say, I will not have this arrow pulled out until
I know by what man I was wounded, whether he is of
the warrior caste, or a brahmin, or of the agricultural
or the lowest caste. Or if he were to say, I will not
have this arrow pulled out until I know of what name
of family the man is;—or whether he is tall, or short,
or of middle height; or whether he is black, or dark, or
yellowish; or whether he comes from such and such a
village, or town, or city; or until I know whether the
bow with which I was wounded was a* chapa *or a*
kodanda, *or until I know whether the bow-string was
of swallow-wort, or bamboo fiber, or sinew, or hemp,
or of milk-sap tree, or until I know whether the shaft
was from a wild or cultivated plant; or whether it was
feathered from a vulture's wing or a heron's or a
hawk's, or a peacock's; or whether it was wrapped
round with the sinew of an ox, or of a buffalo, or of a
ruru-deer, or of a monkey; or until I know whether it
was an ordinary arrow, or a razor-arrow, or an iron
arrow, or of a calf-tooth arrow. Before knowing all
this, verily, that man would have died.*

   *Similarly, it is not on the view that the world is
eternal, that it is finite, that body and soul are distinct,
or that the Buddha exists after death, that a religious
life depends. Whether these views or their opposites
are held, there is still rebirth, there is old age, there is
death, and grief, lamentation, suffering, sorrow, and
despair. . . . I have not spoken to these views because
they do not conduce to absence of passion, or to
tranquillity and Nirvana.*

*And what have I explained? Suffering have I
explained, the cause of suffering, the destruction of
suffering, and the path that leads to the destruction of
suffering have I explained. For this is useful.*[9]

4. Buddha preached a religion devoid of tradition. He
stood on top of the past and its peaks extended his vision
enormously, but he saw his contemporaries as largely buried
beneath those peaks. He encouraged his followers, therefore,
to slip free from the past's burden. "Do not go by what is
handed down, nor on the authority of your traditional teachings. When you know of yourselves: 'These teachings are not
good: these teachings when followed out and put in practice
conduce to loss and suffering'—then reject them."[10] His most
important personal break with archaism lay in his decision—
comparable to Martin Luther's decision to translate the Bible
from Latin into German—to quit Sanskrit and teach in the
vernacular of the people.

5. Buddha preached a religion of intense self-effort. We
have noted the discouragement and defeat that had settled
over the India of Buddha's day. Many had come to accept the
round of birth and rebirth as unending, which was like re-
signing oneself to a sentence of hard labor for eternity. Those
who still clung to the hope of eventual release had resigned
themselves to the brahminic teaching that the process would
take thousands of lifetimes, during which they would gradually work their way into the brahmin caste, the only one
from which release was possible.

Nothing struck the Buddha as more pernicious than this
prevailing fatalism. He denied only one assertion, that of the
"fools" who say there is no action, no deed, no power. "Here
is a path to the end of suffering. Tread it!" Moreover, every
individual must tread this path himself or herself, through

self-arousal and initiative.[11] No god or gods could be counted on, not even the Buddha himself. When I am gone, he told his followers in effect, do not bother to pray to me; for when I am gone I will be really gone. "Buddhas only point the way. Work out your salvation with diligence."[12] The notion that only brahmins could attain enlightenment the Buddha considered ridiculous. Whatever your caste, he told his followers, you can make it in this very lifetime. "Let persons of intelligence come to me, honest, candid, straightforward; I will instruct them, and if they practice as they are taught, they will come to know for themselves and to realize that supreme religion and goal."

6. Buddha preached a religion devoid of the supernatural. He condemned all forms of divination, soothsaying, and forecasting as low arts, and, though he concluded from his own experience that the human mind was capable of powers now referred to as paranormal, he refused to allow his monks to play around with those powers. "By this you shall know that a man is *not* my disciple—that he tries to work a miracle." All appeal to the supernatural and reliance on it amounted, he felt, to looking for shortcuts, easy answers, and simple solutions that could only divert attention from the hard, practical task of self-advance. "It is because I perceive danger in the practice of mystic wonders that I strongly discourage it."

Whether the Buddha's religion—critical of authority, ritual, speculative theology, tradition, reliance on divine aid, and supernaturalism—was also a religion without God will be reserved for later consideration. After his death all the accoutrements the Buddha labored to protect his religion from came tumbling into it, but as long as he lived, he kept them at bay. As a consequence, original Buddhism presents us with a version of religion that is unique and therefore historically invaluable, for every insight into the forms that religion can take increases our understanding of what in essence religion really is. Original Buddhism can be characterized positively as follows:

1. It was empirical. Never has a religion presented its case with such unequivocal appeal to direct validation. On every question personal experience was the final test of truth. "Do not go by reasoning, nor by inferring, nor by argument."[13] A true disciple must "know for him- or herself."

2. It was scientific. Not only did it make the quality of lived experience its final test, it directed its attention to discovering natural cause-and-effect relationships that affected that experience. There is no effect without its cause, and no supernatural beings who interrupt the basic causal processes of the world. The Buddha himself considered his greatest contribution to be the discovery of a *causal law*—dependent arising—whose short version runs, "That being present, this becomes; that not being present, this does not become."[14]

3. It was pragmatic—a transcendental pragmatism if one wishes, to distinguish it from the kind that focuses exclusively on practical problems in everyday life, but pragmatic all the same in being concerned with problem solving. Refusing to be sidetracked by speculative questions, Buddha kept his attention riveted on predicaments that demanded solution. Unless his teachings were useful tools, they had no value whatsoever. He likened them to rafts; they help people cross streams, but are of no further value once the farther shore is reached.

4. It was therapeutic. Pasteur's words, "I do not ask you either your opinions or your religion; but what is your suffering?" could equally have been his. "One thing I teach," said the Buddha, "suffering and the end of suffering. It is just Ill and the ceasing of Ill that I proclaim."[15]

5. It was psychological. The word is used here in contrast to "metaphysical." Instead of beginning with the universe and moving to the place of human beings within it, the Buddha invariably began with the human lot, its problems, and the dynamics of coping with them.

6. It was egalitarian. With a breadth of view unparalleled in his age and infrequent in any, he insisted that women were as capable of enlightenment as men. And he rejected the caste system's assumption that aptitudes were hereditary. Born a *kshatriya* (warrior, ruler), yet finding himself temperamentally a brahmin, he broke caste, opening his order to all, regardless of social status.

7. It was directed to individuals. Buddha was not blind to the social side of human nature. He not only founded a religious community (sangha) that he hoped would become the nucleus of an enlightened society,[16] he insisted on the importance of having "spiritual friends" to reinforce individual resolves: "Noble Friendship is the entire Holy Life."[17] Yet in the end his appeal was to the individual, that each should proceed toward enlightenment through confronting his or her individual situation and predicaments. "Therefore, O Ananda, be lamps unto yourselves. Betake yourselves to no external refuge. Hold fast as a refuge to the Truth. Work out your own salvation with diligence."[18]

~·~·~

THE FOUR NOBLE TRUTHS

When the Buddha finally managed to break through the spell of rapture that rooted him to the Immovable Spot for the forty-nine days of his enlightenment, he arose and began a walk of over one hundred miles toward India's holy city of Banaras. Six miles short of that city, in a deer park at Sarnath, he stopped to preach his first sermon, "Setting-in-Motion the Wheel of the Dharma."[1] The congregation was small—only the five ascetics who had originally shared his severe austerities and then broken with him in anger when he renounced that approach, but who now had become his first disciples. His subject was the Four Noble Truths. His first formal discourse after his awakening, it was a declaration of the key discoveries that had come to him as the climax of his six-year quest.

Asked to list in propositional form their four most considered convictions about life, most people would probably stammer. The Four Noble Truths constitute Buddha's answer to that request. Together they stand as the axioms of his system, the postulates from which the rest of his teachings logically derive.

The First Noble Truth is that life is *dukkha,* usually translated "suffering." Though far from its total meaning, suffering is an important part of that meaning and should be brought into focus before proceeding to other connotations.

Contrary to the view of early Western interpreters, the Buddha's philosophy was not pessimistic. A report of the human scene can be as grim as one pleases; the question of pessimism does not arise until we are told whether it can be improved. Because the Buddha was certain that it could be, his outlook falls within Heinrich Zimmer's observation that "everything in Indian thought supports the basic insight that, fundamentally, all is well. A supreme optimism prevails everywhere." But the Buddha saw clearly that life as typically lived is unfulfilling and filled with insecurity.

He did not doubt that it is possible to have a good time and that having a good time is enjoyable, but two questions obtruded. First, how much of life is thus enjoyable? And second, at what level of our being does such enjoyment proceed? Buddha thought the level was superficial, sufficient perhaps for animals, but leaving deep regions of the human psyche empty and wanting. By this understanding even pleasure is gilded pain. "Earth's sweetest joy is but disguised pain," William Drummond wrote, while Shelley speaks of "that unrest which men miscall delight." Beneath the neon dazzle is darkness; at the core—not of reality, but of unregenerated human life—is the "quiet desperation" Thoreau saw in most people's lives. That is why we seek distractions, for distractions divert us from what lies beneath their surface. Some may be able to distract themselves for long periods, but the darkness is not relieved.

Lo! as the wind is, so is mortal life:
A moan, a sigh, a sob, a storm, a strife.[2]

That such an estimate of life's usual condition is prompted more by realism than by morbidity is suggested by the extent to which thinkers of every stripe have shared it. Existentialists describe life as a "useless passion," "absurd," "too much *(de trop)*." Bertrand Russell, a scientific humanist, found it difficult to see why people should take unhappily to news that the universe is running down, inasmuch as "I do not see how an unpleasant process can be made less so [by being] indefinitely repeated." Poetry, always a sensitive barometer, speaks of "the pitiful confusion of life" and "time's slow contraction on the most hopeful heart." The Buddha never went further than Robert Penn Warren:

Oh, it is real. It is the only real thing.
Pain. So let us name the truth, like men.
We are born to joy that joy may become pain.
We are born to hope that hope may become pain.
We are born to love that love may become pain.
We are born to pain that pain may become more
Pain, and from that inexhaustible superflux
We may give others pain as our prime definition.[3]

Even Albert Schweitzer, who considered India pessimistic, echoed the Buddha's appraisal almost to idiom when he wrote, "Only at quite rare moments have I felt really glad to be alive. I could not but feel with a sympathy full of regret all the pain that I saw around me, not only that of men, but of the whole creation."

Dukkha, then, names the pain that to some degree colors all finite existence. The word's constructive implications come to light when we discover that it was used in Pali to refer to wheels whose axles were off center or bones that had slipped from their sockets. (A modern metaphor might be a shopping cart we try to steer from the wrong end.) The exact meaning of the First Noble Truth is this: Life (in the condition it has got itself into) is dislocated. Something has gone wrong. It is out of joint. As its pivot is not true, friction (interpersonal conflict) is excessive, movement (creativity) is blocked, and it hurts.

Having an analytical mind, the Buddha was not content to leave this First Truth in this generalized form. He went on to pinpoint six moments when life's dislocation becomes glaringly apparent. Rich or poor, average or gifted, all human beings experience:

1. The trauma of birth. Psychoanalysts have in our time made a great deal of this point. Though Freud came to deny that the birth trauma was the source of all later anxiety, to the end he considered it anxiety's prototype. The birth experience "involves just such a concatenation of painful feelings, of discharges and excitation, and of bodily sensations, as have become a prototype for all occasions on which life is endangered, ever after to be reproduced again in us as the dread of 'anxiety' conditions."[4]

2. The pathology of sickness. All bodies fall ill, more or less often, more or less severely, sooner or later.

3. The morbidity of decrepitude. In the early years sheer physical vitality joins with life's novelty to render life almost automatically good. In later years the fears arrive: fear of financial dependence; fear of being unloved and unwanted; fear of protracted illness and pain; fear of being physically repulsive and dependent on others; fear of seeing one's life as a failure.

4. The phobia of death. On the basis of years of clinical practice, Carl Jung reported that he found death to be the

deepest terror in every patient he had analyzed who had passed the age of forty. Existentialists join him in calling attention to the extent to which the fear of death mars healthy living.

5. To be tied to what one dislikes. Sometimes it is possible to break away, but not always. An incurable disease, a stubborn character defect—for better or for worse, there are martyrdoms to which people are chained for life.

6. To be separated from what one loves.

No one denies that the shoe of life pinches in these six places. The First Noble Truth pulls them together with two conclusions, in effect, two deeper, more pervasive dimensions of dukkha. First, even if one *gets* what one loves, the delight doesn't last. All pleasure fades, leaving thirst for renewed pleasure in its wake. In short, *everything*—from the simplest gratifications to the greatest ecstasies—is subject to the universal law of *impermanence* (Skt., *anitya*; Pali, *anicca*). As long as the human heart craves lasting satisfaction, impermanence assures dukkha's presence.

Second, it is not only the grasped-for world of experience that is impermanent; we, the graspers, are as well. The Buddha taught that what we usually think of as our "self" is actually an ever-changing product of five co-conditioning components *(skandhas)*, namely, body, sensations, perceptions, dispositional tendencies, and consciousness. Because *they themselves* are instinctively but ignorantly grasping for a center, a "self" that is not there,[5] they themselves are unsatisfying. "The five groups of grasping are themselves dukkha," says the Buddha.[6] The problem just can't get any deeper. Human beings are caught in a whirlpool of energies they little understand, and until this ignorance is overcome, real happiness is precluded.

For the rift to be healed we need to know its cause, and the Second Noble Truth identifies it. The cause of life's dislocation is *tanha*. Again, imprecisions of translations—all are to some degree inaccurate—make it wise to stay close to the original word. Tanha is usually translated as "desire." There is some truth in this, but if we try to make "desire" tanha's equivalent, we run into difficulties. To begin with, the equivalence would make this Second Truth unhelpful, for to shut down desires, all desires, in our present state would be to die, and to die is not to solve life's problem. But beyond being unhelpful, the claim of equivalence would be flatly wrong, for there are some desires the Buddha explicitly advocated—the desire for liberation, for example, or for the happiness of others.

Tanha is a specific kind of desire, the desire for private fulfillment. When we are selfless, we are free, but that is precisely the difficulty—to maintain that state. Tanha is the force that ruptures it, pulling us back from the freedom-of-all to seek fulfillment in our private egos, which ooze like secret sores. Tanha consists of all

> those inclinations which tend to continue or increase separateness, the separate existence of the subject of desire; in fact, all forms of selfishness, the essence of which is desire for oneself at the expense, if necessary, of others. Life being one, all that tends to separate one aspect from another must cause suffering to the unit which, usually unconsciously, works against the Law. Our duty to our fellows is to understand them as extensions of ourselves—fellow facets of the same Reality.[7]

This is some distance from the way people normally understand their neighbors. The customary human outlook lies a good half way toward Ibsen's description of a lunatic asylum in which "each shuts himself in a cask of self, the cask stopped with a bung of self and seasoned in a well of self." Given a group photograph, whose face does one scan for first? It is a small but telling symptom of the devouring cancer that causes sorrow. Where is the man who is as concerned that no one go hungry as that his own children be fed? Where is the woman who is as concerned that the standard of living for the entire world rise as that her own salary be raised? Here, said the Buddha, is where the trouble lies; this is why we suffer. Instead of linking our faith and love and destiny to the whole, we persist in strapping these to the puny burros of our separate selves, which are certain to stumble and give out eventually. Coddling our individual identities, we lock ourselves inside "our skin-encapsulated egos" (Alan Watts) and seek fulfillment through their intensification and expanse. Fools to suppose that imprisonment can bring release! Can we not see that "'tis the self by which we suffer"? Far from being the door to abundant life, the ego is a strangulated hernia. The more it swells, the tighter it shuts off the free-flowing circulation on which health depends, and the more pain increases.

The Third Noble Truth follows logically from the Second. If the cause of life's dislocation is selfish craving, it ceases when such craving is overcome. If we could be released from the narrow limits of self-interest into the vast expanse of universal life, we would be relieved of our torment.

The Fourth Noble Truth prescribes how the cure can be accomplished. The overcoming of tanha, the way out of our captivity, is through the Eightfold Path.

THE EIGHTFOLD PATH

The Buddha's approach to the problem of life in the Four Noble Truths was essentially that of a physician. He began by examining carefully the symptoms that provoke concern. If everything were going smoothly, so smoothly that we noticed ourselves as little as we normally notice our digestion, there would be nothing to worry about and we would have to attend no further to our way of life. But this is not the case. There is less creativity, more conflict, and more pain than we feel there should be. These symptoms the Buddha summarized in the First Noble Truth with the declaration that life is dukkha, or out of joint. The next step was diagnosis.

Throwing rites and faith to the winds, he asked, practically, what is causing these abnormal symptoms? Where is the seat of the infection? What is always present when suffering is present, and absent when suffering is absent? The answer was given in the Second Noble Truth: The cause of life's dislocation is tanha, the drive for private fulfillment.

What, then, of the prognosis? The Third Noble Truth is hopeful: The disease can be cured by overcoming the egoistic drive for separate existence. This brings us to prescription. How is this overcoming to be accomplished? The Fourth Noble Truth provides the answer. The way to the overcoming of self-seeking is through the Eightfold Path.

The Eightfold Path, then, is a course of treatment. It is not, however, an external treatment, to be accepted passively by the patient as coming from without. It is not treatment by pills, or rituals, or grace. Instead, it is treatment by training. People routinely train for sports and their professions, but, with notable exceptions like Benjamin Franklin, they are inclined to assume that one cannot train for life itself. The Buddha disagreed. He distinguished two ways of living. One—a random, unreflective way, in which the subject is pushed and pulled by impulse and circumstance like a twig in a storm drain—he called "wandering about." The second, the way of intentional living, he called the Path. What he proposed was a series of changes designed to release the individual from ignorance, unwitting impulse, and tanha. It maps a complete course of training; steep grades and dangerous curves are posted, and rest areas indicated. By long and patient discipline, the Eightfold Path intends nothing less than to pick one up where one now is and set one down as a different human being, one who has been cured of crippling disabilities. "Happiness he who seeks may win," the Buddha said, "if he practice."

What is this practice the Buddha is talking about? He breaks it down into eight steps. They are preceded, however, by a preliminary one he does not include in his list, but refers to so often elsewhere that we may assume that he was presupposing it here. This preliminary step is *right association*. No one has recognized more clearly than the Buddha the extent to which we are social animals, influenced at every turn by the "companioned example" of our associates, whose attitudes and values affect us profoundly. Asked how one attains illumination, the Buddha began: "An arouser of faith appears in the world. One associates oneself with such a person." Other injunctions follow, but right association is so basic that it warrants another paragraph.

When a wild elephant is to be tamed and trained, the best way to begin is by yoking it to one that has already been through the process. By contact, the wild one comes to see that the condition it is being led toward is not wholly incompatible with being an elephant—that what is expected of it does not contradict its nature categorically, but heralds a condition that, though startlingly different, is viable. The constant, immediate, and contagious example of its yoke fellow can teach it as nothing else can. Training for the life of the spirit is no different. The transformation facing the untrained is neither smaller than the elephant's nor less demanding. Without visible evidence that success is possible, without a continuous transfusion of courage, discouragement is bound to set in. If (as scientific studies have now shown) anxieties are absorbed from one's associates, may not persistence be assimilated as well? Robert Ingersoll once remarked that had he been God, he would have made health contagious instead of disease; to which an Indian contemporary responded: "When shall we come to recognize that health *is* as contagious as disease, virtue as contagious as vice, cheerfulness as contagious as moroseness?" One of the three things for which we should give thanks every day, according to the Buddha, is the company of the holy. Just as bees cannot make honey unless together, human beings cannot make progress on the Way unless they are supported by a field of confidence and concern that Truth-winners generate.

The Buddha agrees. We should associate with Truth-winners, converse with them, serve them, observe their ways, and imbibe by osmosis their spirit of love and compassion.

> *If you find no one to support you on the*
> *spiritual path, walk alone.*
> *If you see a wise person*
> *who steers you away from the wrong path,*
>   *follow him.*
> *The company of the wise is joyful, like reunion*
> *with one's family.*
> *Therefore, live among the wise,*
> *who are understanding, patient, responsible and noble.*
> *Keep their company like the moon moving among the*
>   *stars.*[1]

With this preliminary step in place we may proceed to the Path's eight steps proper.

1. *Right Views.* A way of life always involves more than beliefs, but it can never bypass them completely, for in addition to being social animals, as was just noted, human beings are also rational animals. Not entirely, to be sure—the Buddha would have been quick to acknowledge this. But life needs guidelines, a map the mind can trust if we are to deploy our energies in the right direction. To return to the elephant for illustration, however great the danger in which it finds itself, it will make no move to escape until it has first assured itself that the track it must tread will bear its weight. Without this conviction it will remain trumpeting in agony in a burning wagon rather than risk a fall. Reason's most vociferous detractors must admit that it plays at least this much of a role in human life. Whether or not it has the power to lure, it clearly holds power of veto. Until

reason is satisfied, an individual cannot proceed in any direction wholeheartedly.

Some intellectual orientation, therefore, is needed if one is to set out other than haphazardly. The Four Noble Truths provide this orientation. Suffering abounds, it is occasioned by the drive for private fulfillment, that drive can be tempered, and the way to temper it is by traveling the Eightfold Path. Thus the Four Noble Truths and the Eightfold Path lock together. The fourth of the Four Noble Truths is the Eightfold Path, and the first step of the Eightfold Path—Right Views—consists of the Four Noble Truths.

2. *Right Intent.* Whereas the first step summoned us to make up our minds as to what life's problem basically is, the second advises us to make up our hearts as to what we really want. Is it really enlightenment, or do our affections swing this way and that, dipping like kites with every current of distraction? If we are to make appreciable headway, persistence is indispensable. People who achieve greatness are almost invariably passionately invested in some one thing. They do a thousand things each day, but behind these stands the one thing they count supreme. When people seek liberation with single-mindedness of this order, they may expect their steps to turn from sliding sandbank scrambles into ground-gripping strides.

3. *Right Speech.* In the next three steps we take hold of the switches that control our lives, beginning with attention to language. Our first task is to become aware of our speech and what it reveals about our character. Instead of starting with a resolve to speak nothing but the truth—a resolve that is likely to prove ineffective at the outset because it is too advanced—we will do well to start further back by trying to notice how many times during the day we deviate from the truth, and to follow this up by asking why we did so. Simi-

larly with uncharitable speech. We begin not by resolving never to speak an unkind word, but by watching our speech to become aware of the motives that prompt unkindness. What flabbiness in our character do we need to protect by this deviation from the truth?

After this first step has been reasonably mastered, we will be ready to try some changes. The ground will have been prepared, for once we become aware of how we do talk, the need for changes will become evident. In what directions should the changes proceed? First, toward veracity. The Buddha approached truth more ontologically than morally; he considered deceit more foolish than evil. It is foolish because it reduces one's being. Why do we deceive? Behind the rationalizations, the motive is almost always fear of revealing to others or to ourselves what we really are. Each time we give in to this "protective tariff," the walls of our egos thicken to further imprison us. To expect that we can dispense with our defenses at a stroke would be unrealistic, but it is possible to become progressively aware of them and recognize the ways in which they hem us in.

The second direction in which our speech should move is toward charity. False witness, idle chatter, gossip, slander, and abuse are to be avoided, not only in their obvious forms, but also in their covert ones. The covert forms—subtle belittling, "accidental" tactlessness, barbed wit—are often more vicious because their motives are veiled.

4. *Right Conduct.* Here, too, the admonition (as the Buddha detailed it in his later discourses) involves a call to understand one's behavior more objectively before trying to improve it. The trainee is to reflect on actions with an eye to the motives that prompted them. How much generosity was involved, and how much self-seeking? As for the direction in which change should proceed, the counsel is again toward

selflessness and charity. These general directives are detailed in the Five Precepts, the Buddhist version of the second, or ethical, half of the Ten Commandments:

Do not kill. Strict Buddhists extend this proscription to animals and are vegetarians.

Do not steal. (Do not take what is not given).

Do not lie. (Do not say what is not so).

Do not be unchaste. For monks and the unmarried, this means continence. For the married it means restraint in proportion to one's interests in, and distance along, the Path.

Do not take intoxicants. Intoxicants cloud the mind. It is reported that an early Russian czar, faced with the decision as to whether to choose Christianity, Islam, or Buddhism for his people, rejected the latter two because both would have required that he give up vodka.

5. *Right Livelihood.* The word "occupation" is well devised, for our work does indeed occupy most of our waking attention. Buddha considered spiritual progress to be impossible if the bulk of one's doings pull against it: "The hand of the dyer is subdued by the dye in which it works." Christianity has agreed. Although he explicitly included the hangman as a role society regrettably requires, Martin Luther disallowed usurers and speculators.

For those who are intent enough on liberation to give their entire lives to the project, right livelihood requires joining the monastic order and subscribing to its discipline. For the layperson, it calls for engaging in occupations that promote life instead of destroying it. Again, the Buddha was not content with generalizing. He named names—the professions of his day he considered incompatible with spiritual seriousness. Some of these are obvious: poison peddler, slave trader, prostitute. Others, if adopted worldwide, would be revolutionary: butcher, brewer, arms maker, tax collector (profiteering was then routine). One on his list continues to be puzzling. Why did the Buddha condemn the occupation of caravan trader?

Although the Buddha's explicit teachings about work were aimed at helping his contemporaries decide between occupations that were conducive to spiritual progress and ones that impeded it, there are Buddhists who suggest that if he were teaching today he would be less concerned with specifics than with the danger of people forgetting that earning a living is life's means, not its end.

The Buddhist tradition groups the third, fourth, and fifth steps under the heading of *sila,* morality, making it clear that moral ineptitude risks not the wrath of a deity, but the retardation of one's own inner development. Trying to progress in Buddhist meditation without cleaning up one's act is like trying to ride off on a horse that is firmly tied to a post. For the Buddha, morality is no mere preliminary to be dropped in the later stages of enlightenment, but rather the meditative life's constant accompaniment and consequence. Virtue is enlightenment's seed *and* fruit.[2]

6. *Right Effort.* The Buddha laid tremendous stress on the will. Reaching the goal requires immense exertion; there are virtues to be developed, passions to be curbed, and destructive mind states to be expunged so compassion and detachment can have a chance. A famous verse of the *Dhammapada* reads, "'He robbed me, he beat me, he abused me'—in the minds of those who think like this, hatred will never cease." But the only way such crippling sentiments can be dispelled,

indeed the only way to shake off fetters of any sort, is by what William James referred to as "the slow dull heave of the will." "Those who follow the Way," said Buddha, "might well follow the example of an ox that marches through the deep mire carrying a heavy load. He is tired, but his steady gaze, looking forward, will never relax until he comes out of the mire, and it is only then he rests. O monks, remember that passion and sin are more than the filthy mire, and that you can escape misery only by earnestly and steadily thinking of the Way."[3] Velleity—a low level of volition, a mere wish not accompanied by effort or action to realize it— won't do.

In discussing right effort, the Buddha later added some afterthoughts about timing and balance. Inexperienced climbers, out to conquer their first major peak, are often impatient with the seemingly absurd saunter at which their veteran guide sets out, but before the day is over his staying pace is vindicated. The Buddha had more confidence in the steady pull than in the quick spurt. When a monk named Sona practiced walking meditation until his feet bled, he began to think of quitting the Path altogether. The Buddha went to him and asked:

> "Tell me, Sona, when in earlier days you lived at home, were you not skilled in playing the lute?"
>
> "Yes, Lord."
>
> "And tell me, Sona, when the strings of your lute were too taut, was your lute tuneful?"
>
> "Certainly not, Lord."
>
> "And when your lute strings were too slack, was your lute tuneful?"
>
> "Certainly not, Lord."
>
> "But when your lute strings were neither too taut

*nor too slack but adjusted to an even pitch, did your
lute then have a tuneful sound?"*

"Certainly, Lord."

*"In the same way, Sona, if effort is applied too
strongly it will lead to restlessness, if too slack it will
lead to lassitude. Therefore, keep your effort
balanced."*[4]

The last two steps of the Eightfold Path represent perhaps
the most distinctive aspect of the Buddha's teaching, namely,
the pivotal importance of the practice of meditation. The
most famous symbol of Buddhism, a person sitting cross-
legged, upright, and relaxed in meditation posture, is perhaps
the most eloquent statement of this basic fact. But if we prefer words to pictures, these few from the Buddhist scholar
Edward Conze will suffice: "Meditational practices constitute the very core of the Buddhist approach to life."[5] The
Buddha's own word for "meditation" was *bhavana,* "mental
development" or "mental cultivation," a task that involves
both right concentration (the eighth step) and right mindfulness (the seventh step). We shall discuss each of these steps
briefly here, exploring them further in Chapter 8.

7. *Right Mindfulness.* No teacher has credited the mind
with more influence over life than the Buddha. The best
loved of all Buddhist texts, the just cited *Dhammapada,*
opens with the words: "All that we are is the result of what
we have thought. Our life is shaped by our mind; we become
what we think. Suffering follows unwholesome thought as
the wheels of a cart follow the oxen that draw it. Joy follows
wholesome thought like a shadow that never leaves."[6] And
with respect to the future there is the saying: "Do you want
to predict your future lives? Examine the condition of your
present mind."

The Buddha counsels such continuous self-examination that it appears daunting, but he thought it necessary because he believed that liberation from unconscious, robotlike existence is achieved only by refined awareness. To this end he insisted that we seek to understand ourselves in depth, seeing everything in our mental and physical states as it really is. If we maintain a steady attention to our moods and thoughts, our actions and our body sensations, we perceive that they incessantly arise and disappear and are in no way permanent parts of us. Right mindfulness aims at witnessing all mental and physical events, including our emotions, without reacting to them, neither condemning some nor holding on to others.

Through right-mindfulness practice, then, we arrive at a number of insights. We begin to see that: (1) every mental and physical state is in flux; none is solid or enduring; (2) habitual clinging to these impermanent states is at the root of much of life's dukkha, and this very insight weakens the habit; and (3) we have little control over our mental states and our physical sensations, and normally little awareness of our reactions. Most important, we begin to realize that there is nobody *behind* the mental or physical events, orchestrating them. When the capacity for mindful attention is refined, it becomes apparent that consciousness itself is not continuous. Like the light from a lightbulb, the on/off is so rapid that consciousness seems to be steady, whereas in fact it is not. With these insights, the belief in a separate self-existent self begins to dissolve and freedom to dawn.[7]

8. *Right Concentration.* Though right concentration is traditionally listed as the eighth step, in many ways it comes before right mindfulness, for in order to undertake mindfulness exercises effectively, one must first learn to focus one's mind. To this end, Buddhism counsels patient, persistent attempts at sustaining one's full attention on a single point, a common

one being simply one's breathing. Initial attempts at concentration are inevitably shredded by distractions; slowly, however, attention becomes sharper, more stable, more sustained. During the first days of some Buddhist meditation trainings, effort may be spent on concentration alone, before moving on to mindfulness exercises. But concentration does not end when mindfulness begins; in fact, they are mutually reinforcing (see Chapter 8).

As was noted earlier, in his later years the Buddha told his disciples that his first intimations of deliverance came to him before he left home, on that day when, still a boy, sitting in the cool shade of an apple tree in deep thought, he found himself caught up into a meditative state (see Chapter 1). It was his first faint foretaste of deliverance, and he said to himself, "This is the way to enlightenment." It was nostalgia for the return and deepening of this experience as much as his disillusionment with the usual rewards of worldly life that led him to his decision to devote his life completely to spiritual adventure. The result, as we have seen, was not simply a new philosophy of life. It was regeneration—transformation into a creature who experienced the world in a new way. Unless we see this, we shall be unequipped to fathom the power of Buddhism in human history. Something happened to the Buddha under that Bo Tree, and something has happened to every Buddhist since who has persevered to the final meditative steps of the Eightfold Path. Like a camera, the mind had been poorly focused, but the adjustment has now been made. The *three poisons*—ignorance, craving, and aversion—begin to evaporate,[8] and we see that things were not as we had supposed. Indeed, suppositions of whatsoever sort begin to vanish, to be replaced by direct perception.

~~~

OTHER CORE BUDDHIST CONCEPTS:
Nirvana, Anatta,
the Three Marks of Existence,
Dependent Arising, and Emptiness

The Buddha's total outlook on life is as difficult to be certain of as that of any personage in history. Part of the problem stems from the fact that, like most ancient teachers, he wrote nothing. There is a gap of almost a century and a half between his spoken words and the first written records, and though memory in those times appears to have been incredibly faithful, a gap of that length is certain to raise questions. A second problem arises from the wealth of material in the texts themselves. Buddha taught for forty-five years, and a staggering corpus has come down to us in one form or another. Although the net result is doubtless a blessing, the sheer quantity of materials is bewildering. His teachings remained remarkably consistent over the years, but it was impossible to say things for many minds and in many ways without creating problems of interpretation. These interpretations constitute the third barrier. By the time texts began to appear, partisan schools had sprung up, some intent on minimizing the Buddha's break with brahmanic Hinduism, others intent on sharpening it. This makes scholars wonder how much in what they are reading is the Buddha's actual thought and how much is partisan interpolation.

Undoubtedly, the most serious obstacle to the recovery of the Buddha's rounded philosophy, however, is his own silence at crucial points. We have seen that his burning concerns were practical and therapeutic, not speculative and theoretical. Instead of debating cosmologies, he wanted to introduce people to a different kind of life. It would be wrong to say that theory did not interest him. His dialogues show that he analyzed certain abstract problems meticulously; that he possessed, indeed, a brilliant metaphysical mind. It was on principle that he resisted philosophy, as someone with a sense of mission might shun hobbies as a waste of time.

His decision makes so much sense that it may seem a betrayal to insert a section like this one, which tries forthrightly to identify—and to some extent define—certain key notions in the Buddha's outlook. In the end, however, the task is unavoidable for the simple reason that metaphysics is unavoidable. Everyone harbors some notions about ultimate questions, and these notions affect interpretations of subsidiary issues. The Buddha was no exception. He refused to initiate philosophical discussions, and only occasionally did he let himself be pried from his "noble silence" to engage in them; but certainly he had views. No one who wishes to understand him can escape the hazardous task of trying to discover what they were.

We may begin with *nirvana,* the word the Buddha used to name life's goal as he saw it. Etymologically it means "to blow out" or "to extinguish," not transitively, but as a fire ceases to draw. Deprived of fuel, the fire goes out, and this is nirvana. From such imagery it has been widely supposed that the extinction to which Buddhism points is total annihilation. If this were so, there would be grounds for the accusation that Buddhism is life-denying and pessimistic. As it is, scholars of the last half century have exploded this view. Nirvana is the highest destiny of the human spirit and its literal meaning is "extinction," but what is to be extinguished are the boundaries of the finite self and the three poisons that feed that self: "The extinction of greed, the extinction of hate, the extinction of delusion: this indeed is called Nirvana."[1]

It does not follow that what is left will be nothing. Negatively, nirvana is the state in which the faggots of private desire have been completely consumed and everything that restricts the boundless life has died. Affirmatively, it is that boundless life itself. Buddha parried every request for a positive description of that boundless state, insisting that it was "incomprehensible, indescribable, inconceivable, unutterable"; after all, after we eliminate every aspect of the only consciousness we have known, how can we speak of what is left?[2] One of Buddha's heirs, Nagasena, preserves this point in the following dialogue. Asked what nirvana is like, Nagasena countered with a question of his own:

> "Is there such a thing as wind?"
> "Yes, revered sir."
> "Please, sir, show the wind by its color or configuration or as thin or thick or long or short."
> "But it is not possible, revered Nagasena, for the wind to be shown; for the wind cannot be grasped in the hand or touched; yet wind exists."
> "If, sir, it is not possible for the wind to be shown, well then, there is no wind."

*"I, revered Nagasena, know that there is wind; I am
convinced of it, but I am not able to show the wind."*
*"Even so, sir, nirvana exists; but it is not possible to
show nirvana."*[3]

Our final ignorance is to imagine that our final destiny is
conceivable. All we can know is that it is a condition beyond
the reach of any psychophysical state still tethered to an "I,"
all such states being, to one degree or another, mirages. The
Buddha would venture only one affirmative characterization.
"Bliss, yes bliss, my friends, is nirvana."

Is nirvana God? When answered in the negative, this question has led to opposite conclusions. Some conclude that since
Buddhism professes no God, it cannot be a religion; others,
that since Buddhism obviously is a religion, religion doesn't
require God. The dispute requires that we take a quick look
at what the word "God" means.

Its meaning is not single, much less simple. Two meanings
must be distinguished for its place in Buddhism to be understood. One meaning of God is that of a personal being who
created the universe by deliberate design and periodically in-
tervenes in its natural causal processes. Defined in this sense,
nirvana is not God. The Buddha did not consider it personal
because personality requires definition, which nirvana ex-
cludes. And though he did not expressly deny creation, he
clearly exempted nirvana from responsibility for it. Finally, the
Buddha left no room for supernatural intervention in the natural causal processes he saw governing the world. If absence of
a personal Creator-God is atheism, Buddhism is atheistic.

There is a second meaning of God, however, which (to distinguish it from the first) has been called the Godhead. The
idea of personality is not part of this concept, which appears
in mystical traditions throughout the world. When the

Buddha declared, "There is, O monks, an Unborn, neither become nor created nor formed. Were there not, there would be no deliverance from the formed, the made, the compounded,"[4] he seemed to be speaking in this tradition. Impressed by similarities between nirvana and the Godhead, Edward Conze has compiled from Buddhist texts a series of attributes that apply to both. We are told that

> *Nirvana is permanent, stable, imperishable,*
> *immovable, ageless, deathless, unborn, and unbecome,*
> *that it is power, bliss and happiness, the secure refuge,*
> *the shelter, and the place of unassailable safety; that it*
> *is the real Truth and the supreme Reality; that it is the*
> *Good, the supreme goal and the one and only*
> *consummation of our life, the eternal, hidden and*
> *incomprehensible Peace.*[5]

We may conclude with Conze that nirvana is not God defined as personal creator, but that it stands sufficiently close to the concept of God as Godhead to warrant the linkage in that sense.[6]

The most startling thing the Buddha said about the human self is that it has no soul. The doctrine of *anatta* (literally, "no-self," i.e., no everlasting, unchanging personal identity) has also caused Buddhism to seem religiously peculiar. But, again, the word must be examined. What was the *atta* (Pali for the Sanskrit *atman,* or "self") that the Buddha denied? At the time it had come to signify (a) a spiritual substance that, in keeping with the dualistic position in Hinduism, (b) retains its separate identity forever.

Buddha denied both these features. His denial of spiritual substance—the soul as homunculus, a ghostly wraith within the body that animates the body and outlasts it—appears to have been the chief point that distinguished his concept of transmigration from prevailing Hindu interpretations. Authentic child of India, the Buddha did not doubt that reincarnation was in some sense a fact, but he was openly critical of the way his brahmin contemporaries interpreted the concept. The crux of his criticism may be gathered from the clearest description he gave of his own view on the subject. He used the image of a flame being passed from candle to candle. As it is difficult to think of the flame on the final candle as being the original flame, the connection would seem to be a causal one, in which influence was transmitted by chain reaction but without a perduring substance.

When to this image of the flame we add the Buddha's acceptance of karma, we have the gist of what he said about transmigration. A summary of his position would run something like this: (1) There is a chain of causes threading each life to those that have led up to it and to those that will follow. Each life is in its present condition because of the way the lives that led up to it were lived. (2) Throughout this causal sequence the will retains at least a small degree of freedom. The lawfulness of things makes the present state the product of prior acts, but within each present moment the will, though deeply influenced, is not completely controlled. People *can* shape their destinies and, in doing so, discover still greater freedom. (3) The two preceding points affirm the causal connectedness of life, but they do not entail that a substance of some sort be transmitted. Ideas, impressions, feelings, streams of consciousness, present moments—these are all that we find, no spiritual substrate. Hume and James were right: If there is an enduring self, subject always, never object, it never shows itself.

An analogy can suggest the Buddha's views of karma and reincarnation in a supporting way. (1) The desires and dislikes that influence the contents of my mind—what I pay attention to and what I ignore—have not appeared by accident; they have definite lineages. In addition to attitudes that I have taken over from my culture, I have formed mental habits. These include cravings of various sorts, tendencies to compare myself with others in pride or envy, and dispositions toward contentment and its opposite, aversion. (2) Although habitual reactions tend to become fixed, I am not bound by my personal history; I can have new ideas and changes of heart. (3) Neither the continuity nor the freedom these two points affirm requires that thoughts or feelings be considered entities—things or mental substances that are transported from mind to mind or from moment to moment. Acquiring a concern for justice from my parents did not mean that a substance, however ethereal and ghostlike, leapt from their heads into mine.

This denial of spiritual substance was only an aspect of Buddha's wider denial of substance of every sort. Substance carries both a general and a specific connotation. Generally, it refers to something relatively permanent that underlies surface changes in the thing in question; specifically, this more basic something is thought to be matter. The psychologist in Buddha rebelled against the latter notion, for to him mind was more basic than matter. The empiricist in him, for its part, challenged the implications of a generalized notion of substance. It is impossible to read much Buddhist literature without catching its sense of the transitoriness (anicca) of everything finite, its recognition of the perpetual perishing of every natural object. It is this that gives Buddhist descriptions of the natural world their poignancy.

Snow falls upon the river,
White for an instant then gone forever.

To what indeed shall I liken
The world and the life of man?
Ah, the reflection of the moon
In the dewdrop
On the beak of the waterfowl.[7]

Though colorfully fragrant
The flower will fall.
Who in this world will live forever?

The Buddha listed impermanence (anicca) as the first of his *three marks of existence*—characteristics that apply to everything in the natural order—the other two being suffering (dukkha) and the absence of independent existence (anatta). Nothing in nature is identical with what it was the moment before; in this the Buddha was close to modern science, which has discovered that the relatively stable objects of the macro world derive from particles that are so ephemeral that they barely exist. To underscore life's fleetingness the Buddha called the components of the human self skandhas— skeins that hang together as loosely as yarn—and the body a "heap," its elements no more solidly assembled than grains in a sandpile. But why did the Buddha belabor a point that may seem obvious? Because, he believed, we are freed from the pain of clutching for permanence only if the acceptance of continual change is driven into our very marrow. Followers of the Buddha know well his advice:

Regard this phantom world
As a star at dawn, a bubble in a stream,
A flash of lightning in a summer cloud,
A flickering lamp—a phantom and a dream.[8]

Given this sense of the radical impermanence of all things finite, we might expect the Buddha's answer to the question "Do human beings survive bodily death?" to be a flat no, but actually his answer was equivocal. Ordinary people, when they die, leave strands of finite desire that can only be realized in other incarnations; in this sense at least these persons live on.[9] But what about the *arhat*,[10] the holy one who has extinguished all such desires; does such a one continue to exist? When a wandering ascetic put this question, the Buddha said:

> *"The word 'reborn' does not apply to him."*
>
> *"Then he is not reborn?"*
>
> *"The term 'not-reborn' does not apply to him."*
>
> *"To each and all of my questions, Gautama, you have replied in the negative. I am at a loss and bewildered."*
>
> *"You ought to be at a loss and bewildered, Vaccha. For this doctrine is profound, recondite, hard to comprehend, rare, excellent, beyond dialectic, subtle, only to be understood by the wise. Let me therefore question you. If there were a fire blazing in front of you, would you know it?"*
>
> *"Yes, Gautama."*
>
> *"If the fire went out, would you know it had gone out?"*
>
> *"Yes."*
>
> *"If now you were asked in what direction the fire had gone, whether to east, west, north, or south, could you give an answer?"*
>
> *"The question is not rightly put, Gautama."*[11]

Whereupon Buddha brought the discussion to a close by pointing out that "in just the same way" the ascetic had not rightly put his question. "Feelings, perceptions, dispositional tendencies, consciousness—everything by which the arhat might be denoted has passed away for him. Profound, measureless, unfathomable is the arhat, even as the mighty ocean; 'reborn' does not apply to him nor 'not-reborn,' nor any combination of such terms."[12]

It contributes to the understanding of this conversation to know that the Indians of that day thought that expiring flames do not really go out, but return to the pure, invisible condition of fire they shared before they visibly appeared. But the real force of the dialogue lies elsewhere. In asking where the fire, conceded to have gone out, had gone, the Buddha was calling attention to the fact that some problems are posed so clumsily by our language as to preclude solution by their very formulation. The question of the illumined person's existence after death is one such case. If the Buddha had said, "Yes, she does live on," his listeners would have assumed the persistence of our present mode of experiencing, which the Buddha did not intend. On the other hand, if he had said, "The enlightened one ceases to exist," his hearers would have assumed that he was consigning such a one to total extinction, which, too, he did not intend.

If we try to form a more detailed picture of the state of nirvana, we shall have to proceed without the Buddha's help, not only because he realized almost to despair how far the condition transcends the power of words, but also because he refused to wheedle his hearers with previews of coming attractions.

Even so, it is possible to form some notion of the goal toward which his Path logically points. We have seen that the Buddha regarded the world as one of lawful order in which events are governed by the pervading law of cause and effect he called dependent arising *(pratitya samutpada)*.[13]

Human actions vectored by ignorant desire tend to yield only more of the same. If the trend is unchecked, the wheel of dependent arising only tightens our bondage. Conversely, however, actions vectored by the intention to win spiritual freedom yield *their* kin. If this trend is cultivated, the wheel of dependent arising starts turning in the *other* direction, uncoiling our bonds. This is the life of awakening. The life of the arhat, then, is one of increasing freedom. The arhat grows in autonomy as her former unskillful habits of mind and body unravel and fall away. In this sense the arhat is increasingly free not only from the passions and worries of the world, but also from its happenings in general. With every growth of inwardness, peace and freedom replace the whirling blades of a Waring blender that the lives of those who are prey to circumstance resemble.

"For a disciple thus freed, in whose heart dwells peace, there is nothing more to do. Steadfast is the mind, gained is deliverance. Ah happy indeed, the arhats! In them no craving is found. The 'I am' conceit is rooted out; confusion's net is burst. Translucent is their mind!"[14]

As long as the arhat remains embodied, her freedom from the particular, the temporal, and the changing cannot be complete. But sever this connection with the arhat's final death, and freedom from the finite will be complete. We cannot say with certainty what the state would be like, but we can venture something. The ultimate end of the Path is a condition in which all identification with the historical experience of the finite self disappears, while experience as such not only remains, but is heightened beyond recognition. As an inconsequential dream vanishes completely on awakening, as the stars go out in deference to the morning sun, so individual awareness is eclipsed in the limitless light of total awareness. Some say, "The dewdrop slips into the shining sea." Others prefer to think of the dewdrop as opening to receive the sea itself.

> *A thousand questions remain, but the Buddha is silent.*
> *Others abide our questions. Thou are free.*
> *We ask and ask; thou smilest and art still.*[15]

What can certainly be said is that spiritual freedom brings largeness of life. The Buddha's disciples sensed that he embodied immeasurably more of reality—and in that sense was more real—than anyone else they knew; and they testified from their own experience that advance along his Path enlarged their lives as well. Their worlds seemed to expand, and with each step they felt themselves more alive than they had been before.

The preceding reflection on the arhat's postmortem state arose out of our definition of anatta as no-soul. That definition is correct, but incomplete, and rounding it out will be the last thing we do in this chapter. For the Buddha also used anatta to characterize things no one ever would have claimed had a soul in the first place. "*Sabbe dhamma anatta,*" the Buddha said. "All things (not only persons) are without-a-self." What does this mean?

The crux of the Buddha's Awakening was the discovery of dependent arising: every thing and every process arises *in dependence* upon countless other things and processes. This is the physicists' field theory incarnate. Nothing exists on its own. Buddhists often convey this insight with the image of Indra's Net, a cosmic web laced with jewels at every intersection. Each jewel reflects the others, together with all the reflections *in* the others. In the deepest analysis, each "jewel" is but the reflection of other reflections. Likewise, every thing and every person in the world, like every jewel in Indra's

Net, because dependently arisen, is *empty-of-own-being* (lacking in self-existence). Empty-of-own-being is the wider meaning of anatta, applicable to the animate and inanimate world alike and less confusing than saying that "things" lack "selves." Dependent arising, anatta, and emptiness-of-own-being are thus three ways of expressing the same insight into the interdependence of all things. This is especially notable since *emptiness* is one of the key concepts of the Mahayana form of Buddhism, which we shall presently discuss.

~~~

THERAVADA AND MAHAYANA:
The Great Divide

Thus far we have been looking at Buddhism as it appears from its earliest records. We turn now to Buddhist history and the record it provides of the variations that can enter a tradition as it seeks to minister to the needs of masses of people and multiple personality types.

When we approach Buddhist history with this interest, what strikes us immediately is that it splits. Religions invariably split. In the West the twelve Hebrew tribes split into Israel and Judah. Christendom split into the Eastern and Western churches, the Western church split into Roman Catholicism and Protestantism, and Protestantism splinters. The same happens in Buddhism. The Buddha dies, and before the century is out the seeds of schism have been sown. One approach to the question of why Buddhism split would be through analyzing the events, personalities, and environments the religion became implicated with in its early centuries. We can cut through all that, however, by saying, simply, that Buddhism divided over the questions that have always divided people.

How many such questions are there? How many questions will divide almost every assemblage of people whether in India, New York, or Madrid? Three come to mind.

First, there is the question of whether people are independent or interdependent. Some people are most aware of their individuality; for them, their freedom and initiative are more important than their ties. The obvious corollary is that they see people as making their own way through life; what each achieves will be largely of his or her own doing. "I was born in the slums, my father was an alcoholic, all of my siblings went to the dogs—don't talk to me about heredity or environment. I got to where I am by myself!" This is one attitude. On the other side of the fence are those for whom life's interconnectedness prevails. To them the separateness of people seems tenuous; they see themselves as supported and vectored by social fields that are as strong as those of physics. Human bodies are separate, of course, but on a deeper level we are joined like icebergs in a common floe. "Send not to ask for whom the bell tolls; it tolls for thee."

A second question concerns the relation in which human beings stand, not this time to their fellows, but to the universe. Is the universe friendly—on the whole helpful toward creatures? Or is it indifferent, if not hostile? Opinions differ. On bookstore shelves we find volumes with titles like *Man Stands Alone,* and next to them *Man Does Not Stand Alone* and *Man Is Not Alone.* Some see history as a thoroughly human project in which humanity raises itself by its own bootstraps or progress doesn't happen. For others it is powered by "a higher power that makes for good."

A third dividing question is: What is the best part of the human self, its head or its heart? A popular parlor game used to revolve around the question, "If you had to choose, would you rather be loved or respected?" It is the same point with a different twist. Classicists rank thoughts above feelings; romantics do the opposite. The former seek wisdom; the latter, if they have to choose, compassion. The distinction probably also relates to William James's contrast between the tough-minded and the tender-minded.

Here are three questions that have probably divided people as long as they have been human and continue to divide them. They divided the early Buddhists. One group took as its motto the Buddha's valedictory, "Be lamps unto yourselves; work out your salvation with diligence." Whatever progress those in this group make is the fruit of wisdom—insight into the cause of suffering and its cure. The other group held that compassion is the more important feature of enlightenment, arguing that to seek enlightenment by oneself and for oneself is a contradiction in terms. For them, human beings are more social than individual, and love is the greatest thing in the world.

Other differences gathered around these fundamental ones. The first group insisted that Buddhism was a full-time job; those who made nirvana their central object would have to give up the world and become monks as the Buddha himself had done. The second group, perhaps because it did not rest all its hopes on self-effort, was less demanding. It held that its outlook was as relevant for the layperson as for the professional; that in its own way it was as applicable in the world as in the monastery. This difference left its imprint on the names of the two outlooks. Both called themselves *yanas*, "rafts" or "ferries," for both claimed to carry people across life's sea to the shores of enlightenment. The second group, however, pointing to its doctrine of cosmic help (grace) and its ampler regard for laypeople, claimed to be Buddhism for the masses and thereby the larger of the two vehicles. Accordingly it preempted the name *Mahayana*, the "Big Raft," *maha* meaning "great," as in Mahatma (the Great Souled)

Gandhi. As this name caught on, the other group came to be known, by default, as *Hinayana,* or the "Little Raft."

Not exactly pleased with this invidious designation, the latter preferred to call its Buddhism *Theravada,* the "Way of the Elders." In doing so it regained the initiative by claiming to represent original Buddhism, the Buddhism taught by Gautama himself. The claim is justified if we confine ourselves to the explicit teachings of the Buddha as they are recorded in the earliest texts, the Pali Canon, for on the whole those texts do support the Theravada position. But this fact has not discouraged the Mahayanists from their counterclaim that it is they who represent the true line of succession. They argue that the Buddha taught more eloquently and profoundly by his life and example than by the words the Pali Canon records. The decisive fact about his life is that he did not remain in nirvana after his enlightenment, but returned to devote his life to others. Because he did not belabor this fact, Theravadins (attending too narrowly to his initial spoken words, the Mahayanists contend) overlook the importance of his "great renunciation," and this causes them to read his mission too narrowly.[1]

We can leave the two schools to their dispute over apostolic succession; our concern is not to judge, but to understand the positions each embodies. The differences that have come out thus far may be summarized by the following pairs of contrasts if we keep in mind that they are not absolute but denote differences in emphasis.

1. For Theravada Buddhism progress is up to the individual; it depends on his or her understanding and resolute application of the will. For Mahayanists the fate of the individual is linked to that of all life. Two lines from John Whittier's "The Meeting" summarize the latter outlook:

He findeth not who seeks his own
The soul is lost that's saved alone.

2. Theravada holds that humanity is on its own in the universe. Gods exist as surely as the sun and moon, but are of no help in winning liberation. Self-reliance is our only recourse.

No one saves us but ourselves,
No one can and no one may;
We ourselves must tread the Path:
Buddhas only show the way.

For Mahayana, in contrast, grace is a fact. We can be at peace because a boundless power draws—or if one prefers, propels—everything to its appointed goal. In the words of a famous Mahayana text, "There is a Buddha in every grain of sand."

3. In Theravada Buddhism the prime attribute of enlightenment is wisdom (bodhi), meaning profound insight into the nature of reality, the causes of anxiety and suffering, and the absence of a separate core of selfhood. From these realizations flow automatically the Four Divine Abidings: lovingkindness, compassion, equanimity, and joy in the happiness and well-being of others. From the Mahayana perspective *karuna* (compassion) cannot be counted on to be an automatic fruit. From the beginning compassion must be given priority over wisdom. Meditation yields a personal power that can be destructive if a person has not deliberately cultivated compassionate concern for others as the motive for arduous discipline. "A guard I would be to them who have no protection," runs a typical Mahayana invocation, "a guide to the voyager, a ship, a well, a spring, a bridge for the seeker of the other shore." The theme has been beautifully elaborated by the poet-saint Shantideva:

May I be a balm to the sick, their healer and servitor
    until sickness comes never again;
May I quench with rains of food and drink the anguish
    of hunger and thirst;
May I be in the famine of the age's end their drink and
    meat;
May I become an unfailing store for the poor, and
    serve them with manifold things for their need.
My own being and my pleasures, all my righteousness
    in the past, present and future, I surrender
    indifferently,
That all creatures may win through to their end.[2]

4. The Theravada sangha (community) has traditionally had a monastic bias, whereas the Mahayana sangha has traditionally sought to avoid spiritually privileging monks over laypeople. Monasteries (and to a lesser extent nunneries) are the spiritual dynamos in lands where Theravada predominates, reminding everyone of a higher truth behind visible reality. Monks and nuns—only partially isolated from society because they are dependent on local people for the one daily meal that is put into their begging bowls—are accorded great respect. This veneration is extended to people who assume monastic vows for limited periods (a not uncommon practice) in order to practice mindfulness meditation intensively. In Burma and Thailand "taking the robe" for a three-month monastic retreat has virtually marked the passage into male adulthood. Mahayana Buddhism, on the contrary, has been primarily a religion for laypeople. Even its priests usually marry, and they are expected to make service to the laity their primary concern.

5. It follows from these differences that the ideal type as projected by the two schools will differ, at least apparently. For Theravadins the ideal is the arhat, the perfected disciple

who, wandering like the lone rhinoceros, strikes out alone for nirvana and, with prodigious concentration, proceeds unswervingly toward that goal. The Mahayana ideal, on the contrary, is the *bodhisattva,* "one whose essence *(sattva)* is perfected wisdom (bodhi)"—a being who, having reached the brink of nirvana, voluntarily renounces that prize and, out of compassion, returns to the world to make nirvana available to others. The bodhisattva deliberately sentences himself (or herself—the best loved of all bodhisattvas is the Goddess of Mercy, in China named Kwan Yin) to agelong servitude in order that others, drawing vicariously on the merit thus accumulated, may enter nirvana first.

The difference between the two types is illustrated by the story of four travelers who, journeying across an immense desert, come upon a compound surrounded with high walls. One of the four determines to find out what is inside. He scales the wall and, on reaching the top, leaps to the other side without a word. The second and third do likewise. When the fourth reaches the top of the wall, she sees below her a lush oasis. Though longing to jump over, she resists the impulse. Thinking of other wayfarers who will be coming that way, she climbs back down to tell them, and subsequently other travelers, of her find. The first three were arhats; the last was a bodhisattva, one who vows not to desert this world "until the grass itself is enlightened."

6. This difference in ideal floods back to color the two schools' estimates of the Buddha himself. For one he was essentially a saint; for the other, a savior. Theravadins revere him as a supreme sage who through his own efforts awakened to the truth and became an incomparable teacher who laid out a path for them to follow. A man among men, his very humanness is the basis for the Theravadins' faith that they, too, have the potential for enlightenment. But the Buddha's direct

personal influence ceased with his *parinirvana* (entrance into nirvana at death). He knows nothing more of this world of becoming and is at perfect peace. The reverence felt by the Mahayanists could not be satisfied with this humanness—extraordinary, to be sure, but human nonetheless. For them the Buddha was a world savior who continues to draw all creatures toward him "by the rays of his jewel hands." The bound, the shackled, the suffering on every plane of existence, galaxy beyond galaxy, worlds beyond worlds, all are drawn toward liberation by the glorious "gift rays" of the Lord. For Theravadins a bodhisattva is a Buddha-in-the-making. For Mahayanists she is a Buddha-maker—one who helps others to become Buddhas.

These differences are the central ones, but several others may be mentioned to fill out the picture. Whereas Theravadins follow their founder in considering metaphysics of dubious worth, Mahayana texts spawned elaborate world pictures replete with many-leveled heavens and hells.[3] For Theravadins, meditation does not involve petitionary prayer to a higher power, whereas the Mahayanists, while maintaining introspective meditation practices, added supplication, petition, and calling on the name of the Buddha for spiritual strength and salvation. Whereas Theravada remained conservative to the point of an almost fundamentalistic adherence to the early Pali texts, Mahayana was liberal in almost every respect. It accepted later texts as equally authoritative, was less strict in interpreting disciplinary rules, and had a higher opinion of the spiritual possibilities of women and the laity in general. Finally, Theravada and Mahayana tend to disagree over whether dualism or nondualism is the most skillful and accurate conception of the Path. We postpone a reflection on this last subtle and thorny issue until Chapter 11, "The Image of the Crossing."

Thus, strangely, the religion that began as a revolt against rites, speculation, grace, and the supernatural ends with all of them back in full force and its founder (who was an atheist as far as a personal God was concerned) transformed into a god himself. We can schematize the differences that divide the two great branches of Buddhism as follows, if we bear in mind that the differences are not absolute:

THERAVADA	MAHAYANA
Human beings are emancipated by self-effort, without supernatural aid.	Human aspirations are supported by divine powers and the grace they bestow.
Key virtue: wisdom.	Key virtue: compassion.[4]
Liberation is difficult work. The monastic vocation is ideal for it.	Liberation is as accessible to laypersons as it is to monks and nuns.
Ideal: the arhat who attains nirvana.	Ideal: the bodhisattva who indefinitely postpones nirvana to care for others.
Buddha a saint, supreme teacher, and inspirer.	Buddha a savior.
Downplays metaphysics.	Elaborates metaphysics.
Downplays ritual.	Emphasizes ritual.
Practice emphasizes meditation and study.	Practice includes elaborate rituals and petitionary prayer.

Of the two schools, the Mahayana was the more populous. Its geography covered more ground in Asia's northeast than Theravada covered in the southeast, and its doctrinal flexibility was able to encompass a wide variety of subschools.

The differences that have occupied us thus far have been doctrinal, but there is an important sociopolitical difference between Theravada and Mahayana as well.[5] Theravada sought to incarnate a feature of the Buddha's teachings that has not thus far been mentioned: his vision of an entire society—a civilization—that was founded, like a tripod, on the monarchy, the monastic community, and the laity, each with responsibilities to the other two and meriting services from them in return. South Asian countries that remain to this day Theravadin—Sri Lanka, Burma, Thailand, and Cambodia—took this political side of the Buddha's message seriously, and remnants of his model are discernible in those lands right down to today. China's interest in Buddhism (which was transmitted to the other lands that were to become Mahayanist: Korea, Japan, and Tibet) bypassed its social dimensions, which included education as well as politics. In East Asian lands Buddhism appears as something of a graft. Buddhist missionaries persuaded the Chinese that they possessed psychological and metaphysical profundities the Chinese sages had not sounded, but Confucius had thought a lot about the social order, and the Chinese were not about to be lectured to on that subject by aliens. So China discounted the political proposals of the Buddha and took from his corpus its psychospiritual components and their cosmic overtones. The world still awaits a history of Buddhism that tells the story of the Theravada-Mahayana divide in terms of the way in which (for geographical and historical reasons) Theravada remained faithful to its founder's vision of a Buddhist civilization, and Mahayana Buddhism became a "religion" that is a part, only, of civilizations whose social foundations had already been laid.[6]

The doctrinal differences between Theravada and Mahayana appear to have softened as the centuries have gone by. Following World War II, two young Germans who were disillusioned with Europe went to Sri Lanka to dedicate their lives to the Buddha's peaceable way. Both became Theravada monks. One, his name changed to Nyanaponika Thera, continued on that path; but the other, while on a sightseeing trip to north India, met some Tibetans, switched to their tradition, and became known in the West as Lama Govinda. Toward the close of Nyanaponika's life a visitor asked him about the different Buddhisms the two friends had espoused. With great serenity and sweetness the aging Theravadin replied: "My friend cited the Bodhisattva Vow as the reason for his switch to Mahayana, but I could not see the force of his argument. For if one were to transcend self-centeredness completely, as the arhat seeks to do, what other than compassion would remain?"

‿‿‿

VIPASSANA:
The Theravadin Way of Insight

There remain two aspects of Theravada Buddhism[1] that could not be adequately addressed in the foregoing survey of the Buddha's original teachings: Theravada's close identification with the body of literature known as the Pali Canon and *vipassana,* Theravada's basic mode of meditation.

The Pali Canon. Theravada's claim to be the school closest to the original teachings of the Buddha grows largely out of its relationship to the literature known as the Pali Canon. Less than a year after the Buddha's death, hundreds of monks gathered near the town of Rajagriha to recite and verify his teachings. At this First Buddhist Council, Ananda, the Buddha's shadow and secretary for virtually all of his forty-five-year teaching career, was the first to speak. Uncorking a prodigious memory, he is said to have delivered verbatim every sermon the Buddha ever gave. The result was the Sutta Pitaka, the "basket" *(pitaka)* of "discourses" *(sutta).* Another monk, Upali, recited the entire code of rules the Buddha had enacted to guide the conduct of monks and nuns. The result was the Vinaya Pitaka, the "basket of discipline." These two "baskets" of teachings were memorized by others and passed down orally for several centuries (writing was often reserved for commercial matters). Some Buddhist communities were content with the two baskets; others created a third, called Abhidhamma,[2] woven of summaries and systematized lists of the teachings found in the discourses. When the entire canon was finally written down (on palm leaves stored in baskets), perhaps around 100 B.C.E., the language used was Pali, a vernacular derivative of Sanskrit. If the Buddha's own language was not Pali, it was probably an early form of the Magadhi dialect (in which the famous rock edicts of King Asoka are written; see Chapter 13), to which Pali bears a close resemblance.

The Pali Canon, therefore, is the earliest written form of the Buddha's teachings, recorded in a language close to the one he spoke. It is not surprising that Theravadins treasure it, nor that in their view the sangha has but two central tasks, the "insight task"—the work of meditation—and the "book task"—learning and preaching informed by the Pali scriptures. "Even if there be a hundred or a thousand monks practicing insight meditation, if there is no learning none will realize the Noble Path," says a Pali commentary.[3] Or as a monk once put it: "To seek liberation through meditation alone, without study, is like trying to climb Mt. Everest with one arm."

Theravada Buddhism reached the island of Sri Lanka within two hundred years of the Buddha's death by way of a missionary venture led by King Asoka's son, Mahinda. The Sri Lankan King Tissa embraced the Buddha's Way, built a large monastery in his capital, Anuradhapura, supported the spread of the teachings over the island, and a few years later welcomed Mahinda's sister, Sanghamitta, to establish a Sri Lankan order of Buddhist nuns.[4] The Theravada reputation for conservatism was born in attempts to distinguish Buddhist teachings from the indigenous beliefs of the island and from the waves of Hindu and Mahayana Buddhist influence that occasionally broke upon its shores.

The Theravada school had also reached Burma around the time it arrived in Sri Lanka and something of a synergy gradually developed. Around the end of tenth century C.E., for example, war in Sri Lanka had extinguished Buddhism, and a contingent of Burmese monks had to be imported to rekindle it. Burmese and Sri Lankan Theravada reinforced each other sufficiently, so that by the time Buddhism died out in India in the eleventh century, it had established a stable home in these countries. Gradually the Theravada form of Buddhism spread to Thailand, Laos, and Cambodia.

The innovations of the Mahayana (Chapter 7) required the writing of new scriptures beyond the Pali canon. Theravadins have always regarded these skeptically; after all the Buddha was long dead and a full record of his teaching was already contained in the Pali Canon. To Theravadin eyes, therefore, Mahayanist creativity threatens to distort the Buddha's teaching into something he never intended. (The Mahayanist reply, of course, is that the Theravadin focus on the letter of the teachings occludes some of their spirit.)

Vipassana Meditation. The form of mental cultivation practiced by Theravada Buddhists has become generally known as *vipassana*, roughly translated as "insight" or "penetrative seeing."[5]

In the *Dhammapada*, the Buddha counsels:

> *As irrigators make water go where they want,*
> *As archers make their arrows straight,*
> *As carpenters carve wood,*
> *The wise shape their minds.*[6]

Thus advised, Theravadins begin the mind shaping work of

vipassana with an effort to increase their capacity for pro-longed concentration. At the outset of a typical training pe-riod, a meditator is asked to fix her attention on a single object, most commonly, the breath. The instruction might sound something like this: "Focus your entire attention on your incoming and outgoing breath. Try to sustain your at-tention there without distraction. If you get distracted, calmly return your attention to the breath and start again." A practitioner might work with this instruction hour upon hour for a number of days.

The Pali term for this concentration training is *samatha,* "tranquillity." The idea is that as one trains one's attention to remain undistractedly aware of a single object, concentration deepens and the mind's usual turbulence dissipates. Initially like the choppy surface of a windblown lake, the mind gradu-ally becomes more like the lake's surface on a windless day.

Samatha, this cultivation of calmness via concentration, is not limited to the Theravada school, but is something of a common denominator among Buddhist meditative traditions. Historian Rick Fields identifies samatha as *the* basic Bud-dhist meditation practice and reports that nearly every lin-eage recommends beginning with some form of it.[7] In the Theravada tradition, however, samatha is not an end in itself, but a gateway to further meditative work. Without some measure of samatha, further steps are impossible. With it, however, two alternative possibilities open up: *jhana* and vipassana. Schematically:

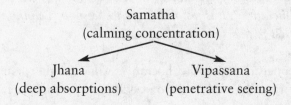

Samatha
(calming concentration)

Jhana
(deep absorptions)

Vipassana
(penetrative seeing)

Jhanas are profound states of absorption that can be in-
duced by calming the mind. The concentration required to
achieve these extraordinary states of mind obviously relates to
the last two steps of the Eightfold Path, but if jhanas are clung
to they can be harmful. Contemporary Theravadin teachers
do not particularly encourage them, and it is widely believed
that the jhanas are neither sufficient nor necessary conditions
for liberation. The already mentioned Nyanaponika Thera,
one of the great Theravadin scholar-monks of the twentieth
century and the author of *The Heart of Buddhist Meditation*,
held this opinion. A few years before he died, he gave this interview:

> Interviewer: *The Pali scriptures recount the Buddha's
> ability to enter jhanas at will and to move between
> the shallowest and the deepest with ease. This
> capacity was also claimed for arhats of the early
> tradition. In your sixty years as a monk in Burma
> and Ceylon, have you ever met anyone with the
> jhanic agility that is described in the canonical
> texts?*
> Nyanaponika: *Well, I met one monk in Burma who
> had reached some jhanic levels, how many I can't
> say. But they are not necessary.*
> Interviewer: *Not necessary?*
> Nyanaponika: *I believe, and my teacher Nyanatiloka
> believed, that even for the final stage of insight
> and the complete purification of the mind [i.e.,
> liberation], it is not necessary to develop even the
> first jhana.*[8]

Skepticism about the liberative value of the jhanas is
partly rooted in the story of Gautama's pre-enlightenment

mastery of the highest jhanas while studying with his first two meditation teachers. When the Buddha-to-be found that these rarified states could *temporarily suspend but not eradicate* the three poisons of ignorance, craving, and aversion, he abandoned them, the scriptures say, "in disgust."[9] The Buddhist commentator Sangharakshita likewise warns that getting "'stuck' in a superconscious state without understanding the necessity of developing insight is for Buddhism not a blessing but an unmitigated disaster."[10]

The most productive use of samatha—again, mind-calming concentration—according to the Theravada tradition, is not as a stepping stone to the jhanas, but as a platform for vipassana, a special kind of self-observation or inner empiricism that produces "penetrative seeing." Before we further describe vipassana, we hasten to add that it builds on samatha but does not replace it; the two work together, reinforcing one another. According to Theravada, it is the *combined* action of concentration (samatha) and insight (vipassana) that discovers the path to freedom.

But what *is* vipassana? If it is "penetrative seeing," what exactly is it looking at? The best way to answer these questions is to turn to the Pali Canon's *Satipatthana Sutta,* "The Discourse on the (Four) Arousings of Mindfulness," which has been described as the most important discourse ever given by the Buddha on mental development. Its opening words are appropriately portentous:

"This is the only way, O bhikkhus, for the purification of beings, for the overcoming of sorrow and lamentation, for the destruction of suffering and grief, for reaching the right path, for the attainment of Nibbana, namely, the Four Arousings of Mindfulness."[11]

In this discourse the Buddha directs a student's samatha-fortified awareness to observe one or more of *four fields*. The first two fields are bodily, the second two mental—though it is axiomatic in Buddhism that "body" and "mind" inhere in and condition one another. Liberation can be worked out only in the laboratory of one's own body and mind, in the immediate actualities of our physical and mental experience. "This I declare, O friend," said the Buddha,

"within this body, six feet long, endowed with perception and cognition, is contained the world, the origin of the world and the end of the world, and the path leading toward the end of the world."[12]

The *first field* for penetrative seeing is simply referred to as *body (kaya)*. Before we cite examples of body awarness contained in the Sutta, it will be helpful to pause a moment to fully appreciate what is going on here. Meditation, viewed from without, is likely to appear exclusively "mental." In meditation, after all, the body, commonly associated with movement, is held motionless. Moreover, because meditation is commonly linked to the vague term "spirituality," it is sometimes tarred with that term's negative connotations toward the body. Those who have not undertaken Buddhist meditation training, therefore, can hardly be expected to guess how intimately connected it is to an awareness of one's own body. The body may be the site of our bondage, but it is also the means of our extrication. Thus, it is not surprising to find the Buddha suggesting that having been born into a human body is one of the three things for which we should give thanks daily.

Indeed, it barely exaggerates the matter to regard Buddhist meditation as a lifelong training in right body awareness. The importance of this point justifies two short quotations. The first is from the "Sayings of the Monks" in the Pali Canon: "If the Buddha granted me a boon and I could obtain that boon, I should choose for all the world, constant awareness directed toward the body."[13] The second is from the Buddha himself:

"One thing, O monks, developed and repeatedly practiced, leads to the attainment of wisdom. It is the contemplation of the body."[14]

The Sutta provides six examples of body mindfulness: (1) *Mindfulness of breathing.* (2) *Mindfulness of basic body postures.* (When you're standing, know it; when you're sitting, know it; when you're lying down, know it; when you're changing from one posture to another, know it.) (3) *Mindfulness of constant change in body activity.* (Know that you are looking now forward, now over your shoulder, now sideways; be aware that you are now bending, now stretching, now sitting this way, now that, now scratching an itch, now chewing, now swallowing, now drinking, now urinating, and so on.) (4) *Mindfulness of the loathsomeness of the body*—a widely misunderstood exercise. The body (as we have seen) is the very laboratory of liberation. Yet the inner work can easily be impeded by the ordinary physical narcissism to which most human beings are subject. The Buddha therefore prescribes an antidote. He says, in effect, imagine how many friends you'd have if your insides were turned out! Understand that the body is in many ways a stinking sack of garbage: hair, nails, pus, bile, blood, phlegm, mucus, sweat, fat, urine, feces, and so on. (5) *Mindfulness of the material elements (earth, air, water, fire) of which the body is composed.* This exercise encourages analytic awareness of body as a field that sometimes feels light, sometimes heavy, sometimes hot, sometimes cold, sometimes relaxed, sometimes tight, and so on. (6) *Mindfulness of the inevitability of the decay of the body.* Recommended is observing corpses rotting at the cremation grounds—which serves the same purpose as the fourth example above.

The *second field* for penetrative seeing is also in the body. But if the body objects in the first field were relatively gross, here in the second field they are subtler. This is the field of *body sensations (vedana)*. The meditator is asked to be closely and steadfastly attentive to the physical sensations that arise at successive moments in various parts of the body, noticing whether they are pleasant, unpleasant, or neutral. The point is not to seek one experience over another, but simply to be aware of—equanimously, that is, without reaction to—what is actually occurring at increasingly subtle levels of body sensation. Working this field one learns indelibly that mental and physical events coproduce one another. As emotions and thoughts arise in the mind, sensations arise in the body, and vice versa. The relatively short space the Sutta gives to this second field—body sensations—is in inverse proportion to its actual importance, for as the Buddha himself put the matter:

"The enlightened one has become liberated by seeing as they really are the arising and passing away of sensations."[15]

The *third field* is *mind*. Just as the first body field concerned general body states, here "mind" refers to general mental states or what might be called moods. The meditator is asked to be steadfastly attentive to whether the mind at the moment is scattered or concentrated, buoyant or heavy, turbulent or calm, worried or confident, confused or clear, and so on. Again, the aim is not to be in this or that state, but simply to notice what mental qualities are in fact present, and that sooner or later they change.

The *fourth field* is *mind objects,* that is, thoughts. Buddhists understand mind as the sixth sense. The mind thinks thoughts the way the body feels feelings; thoughts are, in effect, mind sensations. In illustrating the fourth field, the Sutta does not choose to describe the infinite variety of thoughts that might arise in one's mind; rather, it lists key Buddhist formulas, such as the Four Noble Truths (Chapter 4), the Five Hindrances,[16] the Five Skandhas (Chapter 4), and the Seven Factors of Enlightenment.[17] The point seems to be this: With mindfulness practice well under way, it will help to see the close fit between what one is actually experiencing and the Path's major doctrines. These mind objects therefore provide a kind of analytical deepening and theoretical confirmation of the raw material of one's meditative observation in the other fields.

Theravadin teachers work with these four fields in different ways, but they agree that what is to be penetratively seen again and again, until it sinks into one's marrow, is that all physical and mental events display the three marks of existence—impermanence, lack of self-existence, and unsatisfactoriness. Vipassana, then, is nothing other than seeing with deepening degrees of awareness and equanimity that all phemomena are so marked. The words of the Buddha's disciple Sariputta capture the process:

"Those truths of which before I had only heard, now I dwell having experienced them directly within the body, and I observe them with penetrating insight."[18]

Progressively deeper insight into the three marks within the mental and physical processes that constitute one's "self" is one of the two keys to vipassana's liberating power. The other key lies not in *what* vipassana sees but *how*, namely, *with equanimity*—that is, choicelessly and nonreactively. In the Buddhist view, all of our physical and mental habits—and the "self" they create—harden through repetition and our unthinking identification with them. The practice of motionless sitting makes us acutely aware of the countless and incessant ways these habits clamor and tug for expression. To be steadfastly aware of all these tugs and yet to remain equanimous, not identifying with or reacting to any of them, loosens them up, weakens their hold, and gradually dissolves them. In a word, meditative equanimity deconditions.

As the meditator's old, unskillful modes of conduct erode, new modes of conduct take shape under the influence of the ethical ideals of the Path. One is said to discover new domains of internal freedom, clarity, and fellow feeling. At times, these can even produce a glow that is apparent to others, as here:

> One evening, after the venerable Sariputta had emerged from a day of meditation, he went to the venerable Anuruddha. After he had exchanged with him friendly and polite greetings, he said: "Your features are radiant, friend; your face is bright, your complexion pure. In what abode is your mind dwelling?"
>
> Anuruddha replied: "I often dwell now with mind firmly established in the four foundations of mindfulness."[19]

It has been said that vipassana is present in all Buddhist meditative traditions, but if that is so, it is not always easy to detect.[20] The samatha-vipassana sequence and the cultivation of vipassana within one or more of the four fields of mindfulness seem to be distinctive features of the Theravada tradition of mental cultivation.

There is a final aspect of vipassana meditation that must claim our attention. Clearly its practice requires observing one's own physical and mental processes with a certain amount of detached objectivity. To the casual observer this is bound to appear "unfeeling." Yet the truth is that Buddhist meditation is very much concerned with our feelings and emotions. The Buddha was not out to suppress them, but to refine them. The path of meditation ends not in numbness, but in emotions that produce the purest and most durable forms of human happiness. Meditation, it turns out, is a training in *emotional intelligence*, as the phrase now has it.

Vipassana's emotional-intelligence curriculum is not easy. It requires that one face unflinchingly the bodily ills and emotional pangs that are bound to arise when one locks oneself into the laboratory of the meditation posture. Yet such suffering softens the heart. One sees, perhaps for the first time, that one's ills are not "one's" at all, but belong *necessarily* to every clinging creature in a world that disappoints all clinging. Out of that softened heart, Buddha tells us to *love one another*, to radiate loving-kindness *(metta)* to all sentient beings: "O monks, whatever kinds of worldly merit there are, all are not worth 1/16 part of the heart-deliverance of love (metta); in shining and beaming and radiance, the heart-deliverance of love far excels them."[21] The *Metta Sutta* advises:

> *(Let the skillful one think in this way:)*
> *In safety and in bliss*
> *May all creatures be of a blissful heart.*
> *Whatever breathing beings there may be,*
> *No matter whether they are frail or firm,*
> *With none excepted, be they long or big*
> *Or middle-sized, or be they short or small*
> *Or thick, as well as those seen or unseen,*
> *Or whether they are far or near,*
> *Existing or yet seeking to exist,*
> *May all creatures be of a blissful heart.*
> *Let no one work against another's undoing*
> *Or even slight him at all anywhere;*

And never let them wish each other ill
Through provocation or resentful thought.
And just as might a mother with her life
Protect the son that was her only child,
So let him then for every living thing
Maintain unbounded consciousness in being,
And let him too with love for all the world
Maintain unbounded consciousness in being
Above, below, and all round in between,
Untroubled, with no enemy or foe.
And while he stands or walks, or while he sits
Or while he lies down, free from drowsiness,
Let him resolve upon this mindfulness.[22]

Metta is a Divine Abiding, a heaven on earth. It is not the only one. The Buddha named Four Divine Abidings, four conditions in which human feeling eludes the tether of "I," "me," and "mine" to find fulfillment on an unlimited scale. The other three are compassion (karuna), joy in the good fortune of others *(mudita)*, and equanimity *(upekkha)*, the Buddhist version of the "peace that passeth all understanding." These four states of mind are called boundless because they extend to others feelings that are not bounded by social differences of any sort.

[In] one direction of the world, likewise the second,
likewise the third, likewise the fourth, and so above,
below, around and everywhere, and to all as to
himself, he abides suffusing the entire universe with a
mind of loving-kindness—a mind of compassion—a
mind of sympathetic joy—a mind of equanimity—,
with a mind grown great, lofty, boundless and free
from enmity and ill will.[23]

One way to summarize this chapter is to say that the Theravada's way of insight seeks to clone the aspirant to the full humanity of the Buddha. The Buddha is neither an embodied cosmic principle nor a savior, but a mortal whose own moral courage, meditative discipline, probing intellect, and kind heart can be ours. Two sayings of the Buddha's, above all, capture his legacy:

> *To refrain from evil*
> *To achieve the good,*
> *To purify the mind,*
> *This is the teaching of all Awakened Ones.*[24]

> *Just as the ocean has one taste, the taste of salt, so my Doctrine and Discipline has one taste, the taste of freedom.*[25]

~·~·~

ZEN BUDDHISM:
The Secret of the Flower

In the West the serious study of Buddhism began in Great Britain with Theravada Buddhism and the founding of the Pali Text Society, but in America it began with Zen. This was largely due to the initiative of Paul Carus, who, strongly drawn to Buddhism, virtually adopted a budding young Japanese Buddhist, D. T. Suzuki (of whom we shall hear more in the second part of this book), and brought him to America to translate Buddhist classics while living with the Carus family. Carus founded the Open Court Publishing Company to publish his protégé's prodigious output, and as Suzuki's lineage was Zen, it was primarily in that form that, in the early twentieth century, Buddhism came to America.

Because it is in a primarily Japanese idiom that Zen has taken root in America, that is the idiom in which it will be presented here.[1] Readers should be reminded, however, that Zen is nothing more than the Japanese translation of the Chinese word *Ch'an* and that virtually everything said in this chapter could be said in Chinese, so to speak.

Like other Mahayana sects, Zen claims to trace its ancestry, through China, back to Gautama himself. His teachings that found their way into the Pali Canon, it holds, were true, but provisional. His more perceptive followers heard in his message a subtler teaching, an ultimate truth. The classic instance of this is reported in Buddha's famous "Flower Sermon." Standing on a mountain with his disciples around him, the Buddha did not on this occasion resort to words. He simply held aloft a golden lotus. No one understood the meaning of this eloquent gesture save Mahakasyapa, whose quiet smile, indicating that he had gotten the point, caused the Buddha to designate him as his successor. Legend tells us that the insight that prompted the smile was transmitted in India through twenty-eight patriarchs and carried to China by Bodhidharma in 520 C.E. Spreading from there through Korea to Japan in the twelfth century, it contains the secret of Zen.

Entering Zen is like stepping through Alice's looking glass. One finds oneself in a topsy-turvy wonderland where everything seems quite mad—charmingly mad for the most part, but mad all the same. It is a world of bewildering dialogues, obscure conundrums, stunning paradoxes, flagrant contradictions, and abrupt non sequiturs, all carried off in the most urbane, cheerful, and innocent style imaginable. Here are some examples:

A master, Gutei, whenever he was asked the meaning of Zen, lifted his index finger. That was all. Another kicked a ball. Still another slapped the inquirer.

A Zen master says: The heart is Buddha; this is medicine for sick people. No-heart, no-Buddha: this is for people who are sick from the medicine.

*A novice who makes a respectful allusion to the
Buddha is ordered to rinse his mouth out and never
utter that dirty word again.*

*A twentieth-century Zen master says: There is Buddha
for those who don't understand what he is really.
There is no Buddha for those who understand what he
is really.*

*Someone claiming to understand Buddhism writes the
following stanza:*

> *The body is the Bodhi-Tree;*
> *The mind a mirror bright.*
> *Take heed to keep it always clean,*
> *And let no dust alight.*

*He is at once corrected by an opposite quatrain, which
becomes accepted as the true Zen position:*

> *The Bodhi tree does not exist;*
> *Nor does the mirror bright.*
> *If from the first no-thing exists,*
> *Where can dust alight?*

*A monk approaches a master saying, "I have just come
to this monastery. Would you kindly give me some
instruction?" The master asks, "Have you eaten your
breakfast?" "I have." "Then go wash your bowls." The
monk awakens.*

*A group of Zen masters, gathered for conversation,
have a great time declaring that there is no such thing*

as Buddhism, or enlightenment, or anything even
remotely resembling nirvana. They set traps for one
another, trying to trick someone into an assertion that
might imply the contrary. Practiced as they are, they
invariably elude traps and pitfalls, and each time the
entire company bursts into glorious, room-shaking
laughter.

What goes on here? Is it possible to make any sense out of
what at first blush looks like Olympian horseplay, if not a direct put-on? Can they possibly be serious in this kind of spiritual double-talk, or are they simply pulling our legs?

The answer is that they are completely serious, though it
is true that they are rarely solemn. And though we cannot
hope to convey their perspective completely, it being of Zen's
essence that it cannot be impounded in words, we can give
some hint as to what they are up to.

Let us admit at the outset that even this is going to be difficult, for we shall have to use words to talk about a position
that is acutely aware of their limitations. Words occupy an
ambiguous place in life. They are indispensable to our humanity, for without them we would be howling yahoos. But
they can also deceive, or at least mislead, fabricating a virtual
reality that covers for the one that actually exists. A parent
can be fooled into thinking she loves her child because she ad-
dresses the child in endearing terms. A nation can assume that
the phrase "under God" in its Pledge of Allegiance shows that
its citizens believe in God, when all it really shows is that they
believe in *believing* in God. With all their admitted uses,
words have three limitations. At worst, they construct an arti-
ficial world in which our actual feelings are camouflaged and
people are reduced to stereotypes. Second, even when their descriptions are reasonably accurate, descriptions are not the

things described—the menu is not the meal. Finally, as mystics emphasize, our highest experiences elude words almost entirely.

Every religion that has developed even a modicum of semantic sophistication recognizes to some extent the way words and reason fall short of reality when they do not actually distort it. However much the rationalist may begrudge the fact, paradox and the transrational are religion's lifeblood, as they are the lifeblood of art as well. Mystics in every faith report contacts with a world that startles and transforms them, a world that cannot be plumbed by language. Zen stands squarely in this camp, its only uniqueness being that it makes breaking the language barrier its central concern.

Only if we keep this fact in mind have we a chance of understanding this outlook, which in ways is the strangest expression of mature religion. It was the Buddha himself, according to Zen tradition, who first made the point by refusing in the "Flower Sermon" to equate his experiential discovery with any verbal expression. Bodhidharma continued in this tradition by defining the treasure he was bringing to China as

> *A special transmission outside the scriptures.*
> *With no dependence upon words or letters*
> *A direct seeing into one's own true nature.*

This seems so out of keeping with religion as usually understood as to sound heretical. Think of Hinduism with its Vedas, Confucianism with its Five Classics, Judaism with its Torah, Christianity with its Bible, Islam with its Qur'an. All would happily define themselves as special transmissions *through* their scriptures, and Zen, too, has its texts; they are intoned in its monasteries morning and evening. In addition to the Sutras, which it shares with other branches of Buddhism, it has its own texts: the *Hekigan Roku,* the *Mumonkan,* and others. But one glance at these distinctive texts reveals how unlike other scriptures they are. Almost entirely they are given to pressing home the fact that Zen cannot be equated with any verbal formula whatsoever. Account after account depicts disciples interrogating their masters about Zen, only to receive a roared "Ho!" for an answer. The master sees that through such questions, seekers are trying to fill the lack in their lives with words and concepts instead of realizations. Indeed, students are lucky if they get off with verbal rebuffs. Often a rain of blows is the master's retort. Totally uninterested in the students' physical comfort, the master resorts to the most forceful way he can think of to pry students out of their mental ruts. Those ruts are frighteningly controlling, as Aldous Huxley attests in his one extant limerick:

> *Damn!*
> *I find that I am*
> *A creature that struts*
> *In well-defined ruts*
> *I'm not even a bus, I'm a tram!*

As we might expect, this unique stance toward scripture is replicated in Zen's attitude toward creeds. In contrast to most religions, which pivot around a creed of some sort, Zen refuses to lock itself into a verbal casing; it is "not founded on written words, and [is] *outside* the established teachings," to return to Bodhidharma's putting of the point. Signposts are not the destination, maps not the terrain. Life is too rich and textured to be fitted into pigeonholes, let alone equated with them. No affirmation is more than a finger pointing to the moon. And, lest attention turn to the finger, Zen will point, only to withdraw its finger at once. Other faiths regard blasphemy and disrespect for God's word as sins, but Zen masters may order their disciples to rip their scriptures to shreds and avoid words like "Buddha" or "nirvana" as if they were smut. They intend no disrespect.[2] What they are doing is straining by every means they can think of to blast their novices out of solutions that are only verbal. "Not everyone who says to me, 'Lord, Lord,' will enter the kingdom of heaven" (Matthew 7:21). Zen is not interested in theories about enlightenment; it wants the real thing. So it shouts, buffets, and reprimands without ill will entering in in the slightest. All it wants to do is force students to crash through the word barrier. Minds must be sprung from their verbal bonds into a new mode of apprehending.

Every point can be overstated, so we should not infer from what has been said that Zen forgoes reason and words entirely.[3] To be sure, it is no more impressed with the mind's attempts to mirror ultimate reality than was Kierkegaard with Hegel's metaphysics; no amount of polishing can enable a brick to reflect the sun. But it does not follow that it considers reason worthless. Obviously, it helps us make our way in the everyday world, a fact that leads Zennists in the main to be staunch advocates of education. But more. Working in special ways, reason can actually help awareness toward its goal. If these ways seem like using a thorn to remove a thorn, we should add that reason can also play an interpretive role, serving as a bridge to join a newly discovered world to the world of common sense. There is not a Zen problem whose answer, once discovered, does not make good sense within its own frame of reference; there is no experience that the masters are unwilling to try to describe or explain, given the proper circumstance. The point regarding Zen's relation to reason is a double one. First, Zen logic and description make sense only from experiential perspectives radically different from the ordinary. Second, Zen masters are determined that their students attain the experience itself, not allow talk to take its place.

Nowhere is Zen's determination on this latter point more evident than in the method it adopted for its own perpetuation. Whereas on the tricky matter of succession other religions turned to institutionalized guidelines, papal succession, or creedal dicta, Zen trusted its future to a specific state of consciousness that was to be transmitted directly from one mind to another, like a flame passed from candle to candle or water poured from cup to cup. It is this "transmission of Buddha-mind to Buddha-mind" that constitutes the "special transmission" Bodhidharma claimed was Zen's essence. For a number of centuries this inward transmission was symbolized by the handing down of the Buddha's robe and bowl from patriarch to patriarch, but in the eighth century the Sixth Patriarch in China concluded that even this simple gesture was a step toward confounding outer forms with inner actualities and ordered it discontinued. So here is a tradition that centers in a succession of teachers, each of whom has in principle inherited from his master a mind state analogous to the one Gautama awakened in Mahakasyapa. Practice falls short of this principle, but the following figures suggest the steps that are taken to keep it in place. The master of the teacher under whom one of the authors of this book studied estimated that he had given personal instruction to some nine hundred probationers. Of these, thirteen completed their Zen training, and four were given the *inka* (authorization)—which is to say, they were confirmed as *roshis* (Zen masters) and authorized to teach.

And what is the training by which aspirants are brought toward the Buddha-mind that has been thus preserved? We can approach it by way of three key terms: *zazen, koan,* and *sanzen.*

Zazen literally means "seated meditation." The bulk of Zen training takes place in a large meditation hall. Visitors to these are struck by the seemingly endless hours the monks devote to sitting silently on two long, raised platforms that extend the length of the hall on either side, their faces toward the center (or to the walls, depending on which of the two main lineages of Zen the monastery is attached to).[4] Their position is the lotus posture, adopted from India. Their eyes are half closed as their gaze falls unfocused on the tawny straw mats they are sitting on.

Thus they sit, hour after hour, day after day, year after year, seeking to waken the Buddha-mind so it will suffuse their daily lives. The most intriguing feature of the process is the use they make of one of the strangest devices for spiritual training anywhere to be encountered—the koan.[5] In a general way *koan* means "problem," but the problems Zen devises are fantastic. At first glance they look like nothing so much as a cross between a riddle and a shaggy-dog story. For example:

> *A master, Wu Tsu, says, "Let me take an illustration from a fable. A cow passes by a window. Its head, horns, and the four legs all pass by. Why did not the tail pass by?"*

> *What was the appearance of your face before your ancestors were born?*

> *We are all familiar with the sound of two hands clapping. What is the sound of one hand clapping?* (If you protest that one hand can't clap, you go to the foot of the class.)

Li-ku, a high-ranking officer in the T'ang dynasty, asked a famous Ch'an master: "A long time ago a man kept a goose in a bottle. It grew larger and larger until it could not get out of the bottle anymore. He did not want to break the bottle, nor did he wish to harm the goose. How would you get it out?

The master was silent for a few moments, then shouted, "O Officer!"

"Yes."

"It's out!"

Our impulse is to dismiss these puzzles as absurd, but the Zen practitioner is not permitted to do this. He or she is ordered to direct the full force of the mind upon them, sometimes locking logic with them, sometimes dropping them into the mind's deep interior to wait till an acceptable answer erupts, a project that on a single koan may take as long as a doctoral dissertation.[6] During this time the mind is intently at work, but it is working in a very special way. We in the West rely on reason so fully that we must remind ourselves that in Zen we are dealing with a perspective that is convinced that reason is limited and must be supplemented by another mode of knowing.

For Zen, if reason is not a ball and chain, anchoring mind to earth, it is at least a ladder too short to reach to truth's full heights. It must, therefore, be surpassed, and it is just this surpassing that koans are designed to assist. If they look scandalous to reason, we must remember that Zen is not trying to placate the mundane mind. It intends the opposite: to upset the mind—unbalance it and eventually provoke revolt against the canons that imprison it. But this puts the matter too mildly. By forcing reason to wrestle with what from its normal point of view is flat absurdity, by compelling it to conjoin things that are ordinarily incompatible, Zen tries to drive the mind to a state of agitation wherein it hurls itself against its logical cage with the desperation of a cornered rat. By paradox and non sequiturs, Zen provokes, excites, exasperates, and eventually exhausts the mind until it sees that thinking is never more than thinking *about,* or feeling more than feeling *for.* Then, having gotten the rational mind where it wants it—reduced to an impasse—it counts on a flash of sudden insight to bridge the gap between secondhand and firsthand life.

> *Light breaks on secret lots. . . .*
> *Where logics die*
> *The secret grows through the eye.*[7]

Or again:

> *Ten years of dreams in the forest!*
> *Now on the lake's edge laughing,*
> *Laughing a new laugh.*[8]

Before we dismiss this strange method as completely foreign, it is well to remember that Kierkegaard regarded meditation on the paradox of the Incarnation—the logical absurdity of the Infinite becoming finite, God becoming man—as the most rewarding of all Christian exercises. The koan appears illogical because reason proceeds within structured perimeters. Outside those perimeters the koan is not inconsistent; it has its own logic, a "Riemannian" logic we might say. Once the mental barrier has been broken, it becomes intelligible. Like an alarm clock, it is set to awaken the mind from its dream of rationality. A higher lucidity is at hand.

Struggling with his koan, the Zen monk is not alone. Books will not avail, and koans that are being worked on are not discussed with fellow monks, for this could only produce secondhand answers. Twice a day, though, on average, the monk confronts the master in private "consultation concerning meditation"—sanzen in Rinzai and *dokusan* in the Soto sect. These meetings are invariably brief. The Rinzai trainee states the koan in question and follows it with his or her answer to date. The role of the master is then threefold. In the happy event that the answer is correct, he validates it, but this is his least important role, for a right answer usually comes with a force that is self-validating. A greater service is rendered in rejecting inadequate answers, for nothing so helps the student to put these permanently to one side as the master's categorical rejection of them. This aspect of sanzen is fittingly described in the ninth-century *Rules of Hyakujo* as affording "the opportunity for the teacher to make a close personal examination of the student, to arouse him from his immaturity, to beat down his false conceptions and to rid him of his prejudices, just as the smelter removes the lead and quicksilver from the gold in the smelting-pot, and as the jade-cutter, in polishing the jade, discards every possible flaw."[9] The master's other service is, like that of any exacting examiner, to keep the student energized and determined during the long years the training requires.

And to what do zazen, koan training, and sanzen lead? The first important breakthrough is an intuitive experience called *kensho* ("seeing into one's nature") or *satori* ("understanding"). Though its preparation may take years, the experience itself comes in a flash, exploding like a silent rocket deep within the subject and throwing everything into a new perspective. Fearful of being seduced by words, Zennists waste little breath in describing satoris, but occasionally accounts do appear.

Ztt! I entered. I lost the boundary of my physical body.
I had my skin, of course, but I felt I was standing in
the center of the cosmos. I saw people coming toward
me, but all were the same man. All were myself. I had
never known this world before. I had believed that I
was created, but now I must change my opinion: I was
never created; I was the cosmos. No individual
existed.[10]

From this and similar descriptions we can infer that satori is Zen's version of the mystical experience, which, wherever it appears, brings joy, "at-one-ment," and a sense of reality that defies ordinary language. But whereas most mysticism places such experiences near the end of the religious quest, Zen places them close to the point of departure. In a very real sense Zen training *begins* with satori. For one thing, there must be additional satoris as the trainee learns to move with greater freedom in this realm.[11] But the important point is that Zen, drawing half its inspiration from the practical, commonsense, this-worldly orientation of the Chinese to balance the mystical otherworldly half it derived from India, refuses to permit the human spirit to withdraw—shall we say retreat?—into the mystical state completely. Once we achieve satori, we must

get out of the sticky morass in which we have been
floundering, and return to the unfettered freedom of
the open fields. Some people may say: "If I have
[achieved satori] that is enough. Why should I go
further?" The old masters lashed out at such persons,
calling them "earthworms living in the slime of self-
accredited enlightenment."[12]

The genius of Zen lies in the fact that it neither leaves the world in the less than ideal state in which it finds it nor withdraws from the world in aloofness or indifference. Zen's object is to infuse the temporal *with* the eternal—to widen the doors of perception so that the wonder of the satori experience can flood the everyday world. "What," asks the student, "is the meaning of Bodhidharma's coming from the West?" The master answers, "The cypress tree standing in the garden." Being's amazingness must be directly realized, and satori is its first discernment. But until—through recognizing the interpenetration and convertibility of all phenomena—its wonder spreads to objects as common as the tree in your backyard and you can perform your daily duties with the understanding that each is equally a manifestation of the infinite, Zen's business has not been completed.

With the possible exception of the Buddha himself, in no one is that business ever completely finished. Yet by extrapolating hints in the Zen corpus we can form some idea of what the condition of "the man who has nothing further to do" would be like.

First, it is a condition in which life seems distinctly good. Asked what Zen training leads to, a Western student who had been practicing for seven years in Kyoto answered, "No paranormal experiences that I can detect. But you wake up in the morning and the world seems so beautiful you can hardly stand it."

Along with this sense of life's goodness there comes, second, an objective outlook on one's relation to others; their welfare impresses one as being as important as one's own. Looking at a dollar bill, one's gaze may be possessive; looking at a sunset, it cannot be. Zen attainment is like looking at the sunset. Because Zen practice requires awareness to the full, issues like "Whose awareness?" or "Awareness of what?"

do not arise. Dualisms dissolve. As they do there comes over one three feelings. As one Zen teacher listed them:

> Infinite gratitude to all things past,
> Infinite service to all things present,
> Infinite responsibility to all things future.

Third, the life of Zen (as we have sought to emphasize) does not draw one away from the world; it returns one to the world—the world robed in new light. We are not called to worldly indifference, as if life's object were to spring soul from body as piston from syringe. The call is to discover the satisfaction of full awareness even in its bodily setting. "What is the most miraculous of all miracles?" "That I sit quietly by myself." Simply to see things as they are, as they truly are in themselves, is life enough. It is true that Zen values unity, but it is a unity that is simultaneously empty (because it erases lines that divide) and full (because it replaces those lines with ones that connect). Stated in the form of a Zen algorithm, "All is one, one is none, none is all." Zen wears the air of divine ordinariness: "Have you eaten? Then wash your bowls," we recall. If you cannot find the meaning of life in an act as routine as that of doing the dishes, you will not find it anywhere.

> My daily activities are not different,
> Only I am naturally in harmony with them.
> Taking nothing, renouncing nothing,
> In every circumstance no hindrance, no conflict . . .
> Drawing water, carrying firewood,
> This is supernatural power, this the marvelous
>   activity.[13]

With this perception of the infinite in the finite there comes, fourth, an attitude of generalized agreeableness. "Yesterday was fair, today it is raining"; the experiencer has passed beyond the opposites of preference and rejection. As both pulls are needed to keep the relative world turning, each is welcomed in its proper turn.

A poem by Seng Ts'an, "Trust in the Heart," stands as the purest expression of this ideal of total acceptance:

The perfect way knows no difficulties
Except that it refuses to make preferences;
Only when freed from hate and love
Does it reveal itself fully and without disguise;
A tenth of an inch's difference,
And heaven and earth are set apart.
If you wish to see it before your own eyes
Have no fixed thoughts either for or against it.

To set up what you like against what you dislike—
That is the disease of the mind.
The Way is perfect like unto vast space,
With nothing wanting, nothing superfluous.
It is due to making choices
That its Suchness is lost sight of.

The One is none other than the All, the All none other
*   than the One.*
Take your stand on this, and the rest will follow of its
*   own accord;*
I have spoken, but in vain, for what can words tell
Of things that have no yesterday, tomorrow, or
*   today?*[14]

Even truth and falsity look different. "Do not seek after truth. Merely cease to hold opinions."

Finally, as the dichotomies between self and other, finite and infinite, acceptance and rejection are transcended, even the dichotomy between life and death disappears:

> When this realization is completely achieved, never again can one feel that one's individual death brings an end to life. One has lived from an endless past and will live into an endless future. At this very moment one partakes of Eternal Life—blissful, luminous, pure.[15]

As we leave Zen to its future we may note that its influence on the cultural life of Japan has been enormous. Though its greatest influence has been on pervasive life attitudes, four ingredients of Japanese culture carry its imprint indelibly. In *sumie,* or black-ink landscape painting, Zen monks, living their simple lives close to the earth, have rivaled the skill and depth of feeling of their Chinese masters. In landscape gardening Zen temples surpassed their Chinese counterparts and raised the art to unrivaled perfection. Flower arrangement began in floral offerings to the Buddha, but developed into an art that until recently was a part of the training of every refined Japanese girl. Finally, there is the celebrated tea ceremony, in which an austere but beautiful setting, a few fine pieces of old pottery, a slow, graceful ritual, and a spirit of utter tranquillity combine to epitomize the harmony, respect, clarity, and calm that characterize Zen at its best.

❦❦❦

TIBETAN BUDDHISM:
The Diamond Thunderbolt

To the two paths in Buddhism we have spoken about, we must now add a third. If Theravada (literally, "Elder Way") is the Original Way and Mahayana the Great Way, Vajrayana (a branch of Mahayana) is the Diamond Way.

Vajra was originally the thunderbolt of Indra, the Indian thunder god, who is often mentioned in the early, Pali Buddhist texts; but when Mahayana turned the Buddha into a cosmic figure, Indra's thunderbolt was transformed into the Buddha's diamond scepter. We see here a telling instance of Buddhism's capacity to accommodate itself to local ideas while changing their center of gravity; the diamond transforms the thunderbolt, symbol of nature's power, into an emblem of spiritual supremacy, while retaining the connotations of power that the thunderbolt possesses. The diamond is the hardest stone—one hundred times harder than its closest rival—and at the same time the most transparent stone. This makes the Vajrayana the way of strength and lucidity—strength to realize the Buddha's vision of luminous compassion.[1]

We just noted that the roots of the Vajrayana can be traced back to India, and it continues to survive in Japan as Shingon Buddhism; but it was the Tibetans who perfected this third Buddhist path. Tibetan Buddhism is not just Buddhism with Tibet's pre-Buddhist Bon deities incorporated. Nor is it enough to characterize it as Indian Buddhism in its eighth- and ninth-century heyday moved northward to be preserved against its collapse in India. To catch its distinctiveness we must see it as the third major Buddhist path, while adding immediately that the essence of the Vajrayana is *Tantra*. Tibetan Buddhism, the Buddhism here under review, is at heart Tantric Buddhism.

Buddhists have no monopoly on Tantra, which first showed itself in medieval Hinduism, where the word has two Sanskrit roots. One of these is "extension." In this meaning Tantra denotes texts, many of them esoteric and secret in nature, that were added to the Hindu corpus to extend its range. This gives us only the formal meaning of the word, however. For the content of those extended texts we should look to the second etymological meaning of Tantra, which derives from the weaving craft and denotes "interpenetration." In weaving, the threads of warp and woof intertwine repeatedly. The Tantras are texts that focus on the interrelatedness of things. Hinduism pioneered such texts, but it was Buddhism, particularly Tibetan Buddhism, that gave them pride of place.

The Tibetans say that their religion is nowise distinctive in its goal. What distinguishes its practice is its utilization of all of the energies latent in the human makeup, those of the body emphatically included, and impressing them *all* into the service of the spiritual quest.

The energy that interests—obsesses, one might almost say—the West most is sex, so it is not surprising that Tantra's reputation abroad has been built on its sacramental use of this drive. H. G. Wells once said that God and sex were the only two things that really interested him. If we can have both—not be forced to choose between them as are monks and nuns—this is music to modern ears, so much so that in the popular Western mind Tantra and sex are almost equated. This is unfortunate. Not only does it obscure the larger world of Tantra; it distorts its sexual teachings by removing them from that world.

Within that world Tantra's teachings about sex are neither titillating nor bizarre—they are universal. Sex is so important—after all, it keeps life going—that it must be linked quite directly with God. It is the divine Eros of Hesiod, celebrated in Plato's *Phaedrus* and in some way by every people. Even this, though, is too mild. Sex is the divine in its most available epiphany, but with the proviso that it is such when joined to love. When two people who are passionately, even madly—Plato's divine madness—in love; when each wants most to receive what the other most wants to give; and when at the moment of their mutual climax it is impossible to say whether the experience is more physical or spiritual, or whether the couple sense themselves as two or as one—when these conditions are all in place, we are brought as close as mortals can come to the rapture God enjoys constantly. The moment is ecstatic because at that moment the couple stand outside—*ex*, "out"; *stasis*, "stand"—themselves in the melded oneness of the Absolute.

Nothing thus far is uniquely Tantric; from the Hebrew *Song of Songs* to the explicit sexual symbolism in mystical marriages to Christ, the principles just mentioned turn up in all traditions. What distinguishes Tantra is the way it wholeheartedly espouses sex as a spiritual ally, working with it explicitly and intentionally. Beyond squeamishness and titillation, both, the Tantrics keep the physical and spiritual components of the love-sex splice in strict conjunction— through their art (which shows couples in coital embrace), in their fantasies (the ability to visualize should be actively cultivated), and in overt sexual engagement, for only one of the four Tibetan priestly orders is celibate. Beyond these generalizations it is not easy to go, so we shall leave the matter with a covering observation. Tantric sexual practice is pursued not as a law-breaking revel, but under the careful supervision of a *guru* (spiritual teacher) in the controlled context of a nondualist outlook, and as the climax of a long sequence of spiritual disciplines practiced through many lives. The spiritual emotion that is worked for is ecstatic, egoless, beatific bliss in the realization of transcendent identity. But it is not self-contained, for the ultimate goal of the practice is to descend from the nondual experience better equipped to experience the multiplicity of the world without estrangement.

With Tantra's sexual side thus addressed, we can move on to more general features of its practice. We have already seen that these are distinctive in the extent to which they are body-based, and the physical energies the Tantrics work with most regularly are the ones that are involved with speech, vision, and gestures.

To appreciate the difference in a religious practice that engages these faculties actively, it is useful to think back to Zen. Zazen (seated meditation) sets out to immobilize the body in order to recondition the mind. Tibetans, too, have forms of motionless sitting, but whereas a snapshot could pretty much capture the body's whole repertoire in Zen practice, with the Tibetans a motion picture camera would usually be needed, one that is wired for sound, for, ritualistically engaged, Tibetans' bodies are always moving. Virtually every person who embarks upon the Tantrayana, layperson or monk, is required to perform a foundation (*ngondro,* pronounced nun-dro) practice of one hundred thousand full-body prostrations, sometimes more than once. Tibetan *lamas* (lit., "superior one"—a monk, nun, or advanced lay practitioner) also work with stylized hand gestures, interior visualizations based on their contemplation of sacred art images, and deep-throated chants of sacred sound syllables. Kinesthetically, aurally, and visually, something is always going on.

The rationale they invoke for engaging their bodies in their spiritual pursuits is straightforward. Sounds, sights, and motion *can* distract, they admit, but it does not follow that they *must* do so. It was the genius of the great pioneers of Tantra to discover *upayas* ("skillful means") for channeling physical energies into currents that carry the spirit forward instead of derailing it. The most prominent of these currents relate to the sound, sight, and movement we have referred to, and the names for them all begin with the letter *m. Mantras* convert noise into sound and distracting chatter into holy formulas. *Mudras* choreograph hand gestures, turning them into pantomime and sacred dance. *Mandalas* treat the eyes to icons whose holy beauty draws the beholder in their direction.

If we try to experience our way into the liturgy by which the Tibetans put these Tantric devices into practice, the scene that emerges is something like this. Seated in long, parallel rows, wearing headgear that ranges from crowns to wild shamanic hats, garbed in maroon robes, which they periodically smother in sumptuous vestments of silver, scarlet, and gold, gleaming metaphors for inner states of consciousness, the monks begin to chant. They begin in a deep, guttural, metric monotone, but as the mood deepens those monotones splay out into harmonics that sound like full-throated chords, though actually the monks are not singing in parts; harmony (a Western discovery) is unknown to them. By a vocal device found nowhere else in the world, they reshape their vocal cavities in ways that amplify overtones to the point where they can be heard as discrete tones in their own right.[2] Meanwhile, their hands perform stylized gestures that kinesthetically augment the states of consciousness that are being accessed.

A final, decisive feature of this practice would be lost on observers because it is totally internal. Throughout the exercise the monks visualize the deities they are invoking—visualize them with such intensity (years of practice are required to master the technique) that, initially with closed eyes but eventually with eyes wide open, they are able to see the deities as if they were physically present. This goes a long way toward making them real, but in the meditation's climax, the monks go further. They seek experientially to merge with the gods they have conjured, the better to appropriate their powers and virtues. An extraordinary assemblage of artistic forms are orchestrated here, but not for art's sake. They constitute a technology designed to modulate the human spirit to the wavelengths of the tutelary deities that are invoked.

To complete this profile of Tibetan Buddhism, we must add to this summary of its Tantric practice a unique institution. When in 1989 the Nobel Peace Prize was awarded to His Holiness the Dalai Lama, that institution jumped to worldwide attention.

The Dalai Lama is not accurately likened to the pope, for it is not his prerogative to define doctrine. Even more misleading is the designation God-King, for though temporal and spiritual authority do converge in him, neither of these powers define his essential function. That function is to incarnate on earth the celestial principle of which compassion or mercy is the defining feature. The Dalai Lama is the bodhisattva who in India was known as Avalokiteshvara, in China as the Goddess of Mercy Kwan Yin, and in Japan as Kannon. As Chenrezig (his Tibetan name), he has for the last several centuries incarnated himself for the empowerment and regeneration of the Tibetan tradition. Through his person—a single person who has thus far assumed fourteen successive incarnations—there flows an uninterrupted current of spiritual influence, characteristically compassionate in its flavor. Thus in relation to the world generally, and to Tibet in particular, the office of the Dalai Lama is chiefly neither one of administration nor of teaching, but an "activity of presence" that is operative independently of anything he may, as an individual, choose to do or not do. The Dalai Lama is a receiving station toward which the compassion principle of Buddhism in all its cosmic amplitude is continuously channeled, to radiate thence to the Tibetan people most directly, but by extension to all sentient beings.

Whether the Dalai Lama will reincarnate himself again after his present body is spent is uncertain, for at present the Chinese invaders are determined that there will be no distinct people for him to serve. If there are not, something important will have withdrawn from history, for as rain forests are to the earth's atmosphere, someone has said, so are the Tibetan people to the human spirit in this time of its planetary ordeal.

~·~·~

THE IMAGE OF THE CROSSING

We have looked at three schools of Buddhism: the Theravada and the two Mahayana schools of Zen and Tibetan Buddhism. They are different enough that we must ask whether, on any grounds other than historical lineage, they deserve to be considered aspects of a single religion.

There are six respects in which they should be so regarded. The first five can be propositionalized: (1) They all revere a single founder from whom they claim their teachings derive. (2) They all regard self-centeredness as that which prevents our optimal relation to life. (3) They all regard freedom (from self-centeredness) as a primary theme of the Path. (4) They all emphasize interdependence and our concomitant responsibility to love and care for others. "For hatred can never put an end to hatred. Love alone can. This is an unalterable law."[1] (5) They all adhere to the Buddha's emphasis on impermanence: "People forget that their lives will end soon. For those who remember, quarrels come to an end."[2] Finally, all three can be subsumed under a single metaphor.

This is the image of the crossing, the simple experience of crossing a river on a ferry.

To appreciate the force of this image we must remember the role the ferry played in traditional Asian life. In lands laced by rivers and canals, almost every considerable journey required one. This routine fact underlies and inspires every school of Buddhism. Buddhism is a voyage across life's river, a transport from the commonsense shore of ignorant clinging to the farther bank of compassionate wisdom. Compared with this settled fact, the differences within Buddhism are no more than variations in the type of craft used for the crossing—a raft, a ferry, a boat, whatever.

While we are on the first bank, it is in effect the entire world for us. Its earth underfoot is solid and reassuring. The rewards and disappointments of its social life are vivid and compelling. The opposite shore is barely visible and has no impact on our dealings.

If, however, something prompts us to see what the other side is like, we may decide to attempt a crossing. Making our way to the neighboring dock, we find others who are also waiting for a ferry to arrive. When it comes into view there is an air of excitement. Attention is focused on the distant bank, which is indistinct, but the prospective voyagers are still very much citizens of their familiar world.

The ferry arrives. We clamber aboard, and it pushes out into the water. The bank we are leaving behind is losing its substance. The shops and streets and antlike figures are blending together and releasing their hold on us. Meanwhile, the shore toward which we are headed is not in focus either; it seems almost as far away as it ever was. There is an interval in the crossing when the only tangible realities are the water, with its treacherous currents, and the ferry, which is stoutly but precariously contending with them. This is the moment for Buddhism's Three Vows, also called the Triple Refuge or the Triple Gem: "I take refuge in the Buddha," the fact that there was an explorer who made this trip and proved to us that it is possible. "I take refuge in the Dharma," the map he drew for us. "I take refuge in the Sangha," the company of fellow seekers. The shoreline of our former world has been left behind, and until we set foot on the farther bank, these are the only things in which we can trust.

The yonder shore draws near, becomes real. The ferry jolts onto the sand and we step onto solid ground. This land, which from afar had appeared as misty and unsubstantial as a dream, now provides a sure footing for us. And the shore we left behind, which was then so palpable and real, is now no more than a slender horizontal line, a memory without substance.

In Buddhism this yonder shore is a metaphor for nirvana, the terminus of the Buddhist quest. But this is not to say that nothing more need be done when we set foot on it. The Buddha did not stop meditating after his enlightenment under the Bo Tree, and we can take this as an admonition that for the rest of our lives—and for everyone except the "nonreturners" (those for whom this is their final incarnation) for many lives to come—we, too, must work assiduously toward becoming full citizens of this new country, shedding ourselves incrementally of the encrusted habits we acquired on the side of the world where we formerly lived.

In the image of the crossing, this deepening of our citizenship can be symbolized as pressing deeper and deeper into the interior of this yonder shore. And as we do this, there comes a point when the river behind us and the bank behind *it* disappear from view. All that remains of them now is our memories of them. And from the vantage point that reframes these memories on this "yonder shore" we now see that they are of things that both had to be and were exactly as they should have been in their respective times and places. However unwelcomed they might have been at the time, had we experienced the events in question then as we do now, they would have taken on the wonder and glory of the whole of things for having been required *by* that Whole.

Enter the *Prajnaparamita* (or "Perfection of Wisdom") *Sutras*,[3] which are widely considered to the culminating texts of Mahayana Buddhism. They pivot on a single astounding paradox: "Form is emptiness and emptiness is form; emptiness is not different from form, form is not different from emptiness," where (readers will recall) form refers to conditioned reality and emptiness to nirvana, the unconditioned. The shortest of these Sutras, the *Heart Sutra,* is chanted in Zen monasteries morning and evening, and readers can now recognize the paradoxes, conundrums, and non sequiturs that stud Zen Buddhism as stemming from this root. It collapses the opposites—the river that separated the two shores has disappeared—or, if one prefers, it collapses the opposition between the opposites. It is as radical a nondualism as can be found anywhere, and its consequence is momentous, for it eradicates the disjunction between bondage and freedom. Nirvana is not somewhere else. It is here, now, where the traveler stands; and because this stance happens to be in this world, the world itself is transmuted. The noisy fuss of "liking and disliking" having been stilled, every moment is affirmed for what it actually is. "This our worldly life is coincident with Nirvana itself; not the slightest distinction exists between them," we read.[4] Or as the great Zen master Hakuin put the point:

> This earth on which we stand
> Is the promised Lotus Land,

And this body
Is the body of Buddha.[5]

Such nondualism throws light on the bodhisattva vow not to enter nirvana "until the grass itself be enlightened." As grass keeps coming, does this mean that the bodhisattva will never be enlightened? Not exactly. It means, rather, that his or her "self" has become so transparently interdependent with the All that the idea of a separate "self" attaining an individual enlightenment becomes almost ludicrous. This is already a high degree of enlightenment, but the bodhisattva, claiming no laurels and resting on none, allows compassion to dictate his or her ultimate commitment: "So long as there are beings who do not see what I have seen—that there really are no separate beings and no ultimate difference between *samsara* [the cycle of birth and death] and nirvana—I vow *never* to be finished helping them."[6]

From the standpoint of normal, worldly consciousness, there must always remain an inconsistency between this climactic insight and worldly prudence. This, though, should not surprise us, for it would be flatly contradictory if the world looked exactly the same to those who have crossed the river of ignorance. Only they can dissolve the world's distinctions—or, perhaps we should say, take them in their stride, for the distinctions persist, but now without differences. Where to eagle vision the river can still be seen, it is seen as connecting the two banks rather than dividing them.

THE CONFLUENCE OF BUDDHISM AND HINDUISM IN INDIA

No introduction to Buddhism is complete without mention of the great King Asoka (ca. 272–232 B.C.E.), who was the first to effectively unite the Indian subcontinent. In the history of ancient royalty his figure stands out like a Himalayan peak, clear and resplendent against a sunlit sky. If we are not all Buddhists today it was not Asoka's fault. Not content to simply become a follower of the Buddha himself, he commended the Dharma energetically yet nonviolently across a vast Indian empire stretching to three continents. The speechless creatures of the world have rarely had a more well-connected friend: Asoka outlawed animal sacrifices, ended the slaughter of animals in his kitchens, and renounced royal hunting excursions. His widely disseminated edicts, carved into stone pillars across his empire,[1] dismissed empty ritual and warned against cruelty, anger, and pride while extolling in their place generosity, tolerance, truthfulness, kindness to human and nonhuman animals, and vigor in pursuing the awakened life. Asoka's Buddhist Wheel of the Law still waves on India's flag today. Finding Buddhism an Indian sect, he left it a world religion.

So among the surface paradoxes of Buddhism—a religion that began by rejecting ritual, speculation, grace, mystery, and a personal God and then sprouted a branch that brought them all back into the picture—there is a final paradox. Today Buddhists abound in every Asian land except India; only recently, after a thousand-year absence, are they beginning in small numbers to reappear. Buddhism triumphs in the world at large, only (it would seem) to fail in the land of its birth.

This surface appearance is deceptive. The deeper fact is that in India Buddhism was not so much defeated by Hinduism as accommodated within it. Up to around the year 1000, Buddhism persisted in India as a distinct religion. To say that the Muslim invaders then wiped it out will not do, for Hinduism survived. The fact is that in the course of its fifteen hundred years in India, Buddhism's differences with Hinduism softened. Hindus admitted the legitimacy of many of the Buddha's reforms, and in imitation of the Buddhist sangha orders of Hindu *sadhus* (wandering ascetics) came into existence. From the other side, as Buddhism opened into the Mahayana, its teachings came to sound increasingly like Hindu ones until in the end Buddhism sank back into the source from which it had sprung.

Only if one assumes that Buddhist principles left no mark on subsequent Hinduism can the merger be considered a Buddhist defeat. Actually, almost all of Buddhism's affirmative doctrines found their place or parallel. Its contributions, accepted by Hindus in principle if not always practice, included a renewed emphasis on kindness to all living things, on nonkilling of animals (most upper-class Hindus are vegetarians), and on the elimination of caste barriers in matters religious and their reduction in matters social as well as a strong ethical emphasis generally. The bodhisattva ideal seems to have left its mark in prayers like the following by Rantideva in the great Hindu devotional classic the *Bhagavatam:*

> *I do not pray to the Lord for a state in which I shall be endowed with the eightfold powers, nor even for the state of liberation from the cycle of birth and death. I pray only that I may feel the pain of others, as if I were residing within their bodies, and that I may have the power of relieving their pain and making them happy.*[2]

All in all, the Buddha was more than reclaimed as "a rebel child of Hinduism." He was raised to the status of the ninth[3] of the ten[4] divine incarnations of Vishnu, God in his role of Preserver.

PART II

~~~

THE WHEEL ROLLS WEST

I3

~~~

THE NEW MIGRATION

*Go your ways, O monks, for the benefit of the many,
for the happiness of many, out of compassion for the
world, for the good, benefit, and happiness of gods
and men. No two should go in the same direction.
Teach the Dharma which is beneficent in the
beginning, in the middle, and in the end—both the
spirit and the letter of it. Make known your own pure
way of life.*[1]

Thus urged by the Buddha himself and seconded two centuries
later by King Asoka, Buddhist pilgrims carried their ideas and
ways of life westward out of India. Asokan edicts record the
sending of dharma envoys to Syria, Egypt, and Macedonia,
though there is no mention in Western sources of their arrival.
Some believe that Buddhist monks made it all the way to the
Mediterranean Sea, influencing Greek philosophy and early
Christianity en route, but there is no hard evidence for this.

There is of course plenty of evidence for Buddhism's migration eastward—into China by the first century C.E., into Japan via Korea by the twelfth—but push east far enough and it becomes West. Ancient Chinese documents tell of a far journey undertaken by the Buddhist monk Hui Shan and five fellow monks around 458 C.E. Some scholars believe the six sailed to Alaska, then trekked down America's west coast all the way to Mexico, influencing the culture, and perhaps the DNA, of the *Hui*chol Indians, whose facial features, some say, closely resemble those of the Chinese. If true, this would be the earliest contact known between the Old and New Worlds, predating Columbus by a millennium. But Hui Shan's journey to the West is today still a debated claim rather than an established fact.

One of the first Europeans to make documented contact with Buddhism was the thirteenth-century adventurer Marco Polo. Based on stories he had heard in Mongolia and Ceylon (Sri Lanka), Polo wrote that "for a certainty, if he [the Buddha] had been baptized a Christian he would have been a great saint before God."[2] A century later Jean Marignolli, an envoy to China for Pope Benedict XII, visited Sri Lanka on the way home. He favorably compared the Buddhist monks he met with the Catholic Franciscans, praised their hospitality, and noted, "These people lead a very saintly life." Yet these remarks rank among the kindest things poorly informed Europeans would say about Buddhism for centuries. More common were the slurs and dismissals that fell from the lips of the traders, military men, priests, and missionaries who reached China, Japan, Sri Lanka, Indochina, Tibet, and India after Vasco da Gama's voyage to India in 1497. Buddhism was sometimes called a "monstrous religion," an "abominable sect," and a "plague" initiated by "a very wicked man."[3]

When the Jesuit priest Francis Xavier arrived in Japan in

1549, the tide turned a little. Xavier was at least convinced of the need to more fully understand those he would convert. He realized at once that he was in the midst of developed culture. Among the non-Christian peoples so far discovered by the Christian world, "no other," he wrote, "will be found to surpass the Japanese. In their culture, their social usage, and their mores they surpass the Spaniards so greatly that one must be ashamed to say so."[4] Xavier was befriended by a Zen abbot, and though the latter never converted, the two grew to be warm, mutually respectful friends. Still, Xavier was uneasy with the apparent Buddhist denial of an eternal soul; to his dismay, during discussions of the soul's immortality with his Zen friend, the latter would sometimes say yes and sometimes no. And Xavier was appalled by the Buddhist indifference to a personal Creator God. Indeed, some Japanese, no doubt irked by Xavier's proselytizing, twisted *Deus,* Xavier's Latin term for God, into *Daiuso,* meaning "Great Lie." Still, within fifty years of Xavier's pioneering mission, two hundred thousand Japanese had become Catholic Christians.

Over the next century other Jesuit missionaries to China and Tibet followed Xavier's lead, seeking to convert Buddhists, but in any case slowly adding to European awareness of the tradition. In 1678, Louis XIV's envoy to Siam (Thailand) brought the word "nirvana" to European ears, probably for the first time. In 1727, A *History of Japan* was published in London, the first book in English to describe Zen Buddhism, and it did so sympathetically. At a time when Europe was still reeling from the bloody spectacle of Protestant-Catholic barbarism that was the Thirty Years' War (1618–48), such accounts, as well as the upbeat chronicles of European merchants that told of civilized Asian societies, helped to erode the prejudice that Christian Europe was the sole repository of wisdom and high culture. Interest in the collection, translation, and study of Oriental texts grew for some scholars into a passion, as it did, for example, for William Jones.

An Englishman born in 1736, Jones was destined for a career in law, but his early years were distinguished by the ease with which he learned foreign tongues. Before leaving Harrow in his teens, he read in Latin, Greek, French, Spanish, and Hebrew. At Oxford, he plunged into Arabic, Persian, and Turkish. In his late forties, his legal training and linguistic talents won him a high-court judgeship in India. There the indefatigable Jones embraced the study of Sanskrit and, with other like minds, founded the Asiatick Society, dedicated to furthering knowledge of all things Asian. *Asiatick Researches,* the society's journal, began to collect, among other things, firsthand reports of Buddhism from Sri Lanka and Tibet and to give eighteenth-century European intellectuals their first significant Asian education.[5]

If the seventeenth century had witnessed the Thirty Years' War, it had also (by bringing forth the likes of Galileo and Newton) demonstrated the awesome potential of the scientific method. One of the ideals of the French and German Enlightenment was to construct a "science of humanity" on the same basis—the rational analysis of empirical facts—that was proving so successful in the natural sciences. A truly global science of humanity could, of course, be constructed only from truly global data. As European scholars groped to assemble an understanding of Asian culture via the puzzle pieces made available through journals like *Asiatick Researches,* the contours of Buddhism as a whole were not at first perceived. Not until 1844 did the French philologist Eugene Burnouf demonstrate in a work of stunning scholarship that religious data discovered in Southeast Asia, Tibet,

China, and Japan belonged to branches of a single multiform tradition—Buddhism—that originated in India. Buddhology, the academic study of Buddhism, had been launched. In many ways, this was the key that finally flung wide the door between Buddhism and the West. As Buddhist historian Rick Fields has aptly said, "Until the later nineteenth century, some twenty-four hundred years after Shakyamuni's enlightenment under the bodhi tree, we can find virtually no real exchange between the European mind and Buddhism."[6]

The German philosopher Arthur Schopenhauer (1788–1860) was perhaps the first Westerner to publicly confess a deep affinity with Buddhism. Schopenhauer believed he saw reflections of the Buddha's teaching in his own convictions: that a blind and relentless will-to-live beats in the blood of all life; that freedom from its clutches is the noblest aspiration of the human spirit; and that compassion for fellow creatures, human and nonhuman alike, prepares us for, and offers us a foretaste of, that freedom. Schopenhauer in turn influenced Friedrich Nietzsche (1844–1900), who, while not embracing Buddhism, says intriguing and mostly sympathetic things about it in passages scattered among his tremendously influential books.[7] Religions, in Nietzsche's view, make a thousand promises but keep none, while "Buddhism makes no promises but keeps them [all]."[8]

At this point we must split the story of Buddhism's western journey in two. One strand stays in western Europe, the other leads to North America. The stories are similar, however. We shall merely sketch the European picture in a few suggestive strokes, referring interested readers to resources for further information. We shall then take up the American strand in greater detail in subsequent chapters.

Germany, England, France, and Switzerland are home to about one million Buddhists, 25 percent of whom are European converts, and over 900 Buddhist groups and centers. Buddhists make up roughly 1 percent of the total population of these countries.[9]

Germany. Three years after Nietzsche's death, in 1903, Anton Gueth became the first German national to formally enter the Buddhist order of monks. He did so in Yangon, Myanmar (Rangoon, Burma), taking the name Nyanatiloka. He soon moved to Sri Lanka and founded the Island Hermitage in Dodanduwa, which quickly became known as a Buddhist training center for Europeans. Fifty years later he would write:

> *It was in 1903 that I came first to this island which since then I have considered my spiritual home. Yet it was the great wish of my heart to give the country of my origin the best I possessed, that is, the Dhamma [Dharma]. And to that end I have devoted the greatest part of my 50 years in the Sangha. I did so in the firm conviction that the Dhamma will take root in my home country, Germany, and may have a great future there.*[10]

Nyanatiloka's influential writings and translations[11] soon attracted others to the Path, such as the German Jew Siegmund Feniger, who left Germany in 1936 to become a lifelong Theravadin monk in Sri Lanka under the name Nyanaponika Thera. His own deeply informed writings[12] have made a measureless contribution to the twentieth-century West's knowledge of Buddhism.

Ernst Hoffman traveled a similar trajectory. He became a Theravadin monk, but later switched his allegiance to Tibetan Vajrayana. The West knows of his lasting influence on Western Buddhism under the name Lama Anagarika Govinda.

In 1921, Karl Seidenstucker (1876–1936), who almost

twenty years earlier had founded the pioneering Buddhist Mission in Germany, now cofounded with Judge Georg Grimm (1868–1945) the Ancient Buddhist Community in Munich. Grimm's magnum opus, *The Doctrine of the Buddha,* first published in 1915, and Hermann Oldenberg's classic, *The Buddha* (1881), are early German contributions to the Western understanding of the Dharma that can still be read with great benefit today.

The medical doctor Paul Dahlke (1865–1928) studied Pali Buddhism, wrote books and translations enriched by many visits to Sri Lanka, and was about to become a monk there when ill health made that impossible. Returning to Germany in 1924, he established the Buddhist House in Berlin-Frohnau, which became a recognized center for German Buddhism. Under the guidance of the Sri Lankan German Dhammadhatu Society, the Buddhist House has evolved into the Berlin Buddhist Vihara for the training of monks.

In the first half of the past century, German Buddhism was dominated by Theravadin influence. Interest in Zen picked up after the war. Several Japanese Zen-influenced philosophers were attracted by the philosophy of Martin Heidegger because the latter's late writings sometimes appear to suggest that the twenty-five-hundred-year arc of Western philosophy finds culmination and closure in a Buddhist kind of contemplative awareness. And the Jesuit mission to Japan by Francis Xavier in the sixteenth century eventually bore an unexpected kind of fruit when a German Jesuit, Father Hugo Enomiya-LaSalle (1898–1990), underwent Zen training, became recognized as a master in 1978, and later, while still embracing the Catholic faith, led Zen retreats in Germany and elsewhere.

The German Buddhist Union (DBU) was founded in 1955 by forty-three German Buddhists and now comprises fifty-two multibranched Buddhist groups and centers, a fivefold increase in the last sixteen years. The DBU describes itself as "a platform where Buddhists from all traditions and schools meet and develop understanding and appreciation for the variety and diversity of Buddhist theory and practice." German Buddhists are the only ones of which we are aware who have developed a Buddhist Confession *(Buddhistisches Bekenntnis)* acceptable to a wide spectrum of traditions and schools. Its five points affirm: (1) the Triple Refuge (Buddha, Dharma, and Sangha), (2) the Four Noble Truths, (3) the expression of these principles in an ethical life, (4) which is guided by the five basic moral precepts (see Chapter 5), and (5) the development of the four sublime states of loving-kindness, compassion, sympathetic joy, and equanimity (see Chapter 8). In 1962 there were only 2,000 Buddhists registered in the German Buddhist Union. Today there are about 150,000 Buddhists in Germany, 40,000 of whom are Europeans.

Even so brief a survey should not omit mention of the Swiss-German writer and Nobel laureate Hermann Hesse, who stirred up considerable interest in Buddhism with the publication of *Siddhartha* in 1922. A novella that has won a vast readership in many languages, *Siddhartha* tells of a young man who opts against becoming the Buddha's disciple, chooses instead his own way through life, and learns Buddhist lessons the hard way.

England. An important pioneer for British and Western Buddhism was Thomas Rhys-Davids, who had traveled to Sri Lanka in the late nineteenth century as a British civil servant. Like his countryman of the previous century, William Jones, he soon found that his comprehension of local traditional law required mastery of an ancient classical language, in this case, Pali. Rhys-Davids's Pali teacher happened to be a

Buddhist monk, and his studies soon developed into the first systematic effort to collect the entire Pali Canon, considered by Theravadins to be the earliest and most accurate record of the Buddha's teaching. In England in 1881, Rhys-Davids formed the still extant Pali Text Society to complete the collection, transliterate it from the ancient Pali script into Roman characters, and study its content.

One of the first Britons to take formal refuge in Buddhism was Allan Bennet, who was inspired to do so by Edwin Arnold's poem *The Light of Asia* (see Chapter 14). He was ordained in Burma around 1908. Though ill health forced him to disrobe in 1914, he continued to propagate the Dharma and founded the Buddhist Lodge—whose leading light for years was the barrister and Buddhist author Christmas Humphreys. The lodge continues to this day as the London Buddhist Society UK.

Osbert Moore acquired his *bodhicitta* ("intention toward enlightenment") through reading a book on Buddhism while serving as a British soldier in Italy during World War II. After the war he traveled to Sri Lanka, trained at Nyanatiloka's Island Hermitage, and was ordained in 1950 as Bhikku Nyanamoli. He produced important English translations of Theravada texts before a premature death.

William Purfurst was ordained in Thailand around 1954 as Venerable Kapilavuddho. He founded the English Sangha Trust in 1955. In 1974, the Sangha Trust invited a Buddhist meditation master from Thailand, Ajahn Chah, to come to England to establish an order of Western Buddhist monks. Chah arrived with a surprise, an American, Robert Jackman, who had already been studying with Chah in Thailand for ten years. When Ajaan Chah returned to Thailand, Jackman, now Achaan Sumehdo, was left in charge of the fledgling monastic experiment. It began modestly enough—he and three brother monks lived in cramped quarters in a noisy section of north London. But on an alms round one morning they caught the attention of a jogger who happened recently to have bought an entire forest in West Sussex with the intention of revitalizing it, but with no clear sense of how to do it. Realizing that the monks would make excellent forest dwellers, he simply gave it to them. The monks moved to a house in the nearby village of Chithurst and began to rehabilitate both the house and the forest. This was the beginning of now well-established and lovely Chithurst Forest Monastery. In 1993, Buddhist writer Stephen Batchelor described it as follows:

> *Chithurst is no longer an eccentric outpost of Thai Buddhism in a corner of England, but the nucleus of a growing monastic community throughout Europe. It has become a Buddhist "seminary," where newly ordained monks are trained before being sent to serve in Amaravati, the main public center north of London, or in the two smaller centres in Britain, or even to the newly opened viharas [monasteries] in Switzerland and Italy. There are around fifty monks, nine nuns and about thirty male and female postulants living in the different European centers.*[13]

Since Batchelor wrote this, Amaravati has founded branches in New Zealand, Australia, and America as well.

Lectures by the Sri Lankan monk Narada Thera, in 1949, led to the founding of the London Buddhist Vihara in Knightsbridge in 1954, whose leader between 1957 and 1990 was the scholar Dr. Saddhatissa.

The Burmese Theravada influence began with the founding of a vihara in Birmingham in 1978. There is now a Burmese vihara in London and a meditation center in Billinge in Lancashire. Meditation teachings in these places draw from the Mahasi Sayadaw and U Ba Khin/S. N. Goenka lineages (discussed in Chapter 18).

Even the briefest sketch of British Buddhism cannot afford to omit Dennis Lingwood. Lingwood was born in 1925. By his mid-teens, having read Madame Blavatsky's *Isis Unveiled* and the *Platform* and *Diamond Sutras* of Buddhism, he recognized himself as a Buddhist. Drafted by the British army in 1943, Lingwood was posted first in India, then Sri Lanka. At first he was repelled by the institutional forms of Buddhism he encountered, but his interest in the Dharma continued to grow. Shortly after the war and still in India, Lingwood took formal Theravada ordination and the name Sangharakshita. In the early 1950s, he met and was deeply impressed by the already mentioned German convert to Tibetan Buddhism, Lama Govinda. At the age of twenty-nine he gave a series of lectures on Buddhism in Bangalore. Inspired by their warm reception, he reworked them into his magisterial *Survey of Buddhism*, the first major item in what has become a prodigious output. Convinced of both the need for serious, structured Western commitment to the Buddhist way of life and the lack of full fit between Western society and traditional Buddhist monasticism, Sangharakshita went about founding, in 1967, the Friends of the Western Buddhist Order (FWBO) and, a year later, ordained the first nine men and three women to serve as its monastic nucleus, the Western Buddhist Order (WBO). Today the FWBO is an international network of over one hundred Buddhist centers, retreat facilities, residential communities, right-livelihood businesses, and educational, health, and art programs.

Between 1979 and 1991 the number of British Buddhist groups has tripled. In Britain, out of 370 meditation-oriented

Buddhist groups and organizations, 57 are affiliated with the Zen tradition, 65 with Tibetan Buddhism, and some 100 with Theravada and its central vipassana practice. In Great Britain there are about 150,000 Buddhists, about a third of them Europeans.

France. There are about 650,000 Buddhists in France, roughly 25 percent of whom are Europeans. France has the largest per capita population of Buddhist Europe, 1.15 percent. All three of Buddhism's major extant branches, Theravada, Zen, and Tibetan, are represented. Perhaps the most conspicuous presences are those of the Kagyu Tibetan lineage and the school of Soto Zen.

Of the approximately fifteen Kagyu bases, the largest is Kagyu Ling in Burgundy, founded by Kalu Rinpoche in 1976. One of Kalu's first orders of business was to "close the doors" behind a group of seven men and six women about to undertake the famous Tibetan Buddhist three-year, three-month, and three-day solitary retreat and intensive meditation practice (see Chapter 17, note 8). When the doors opened thirty-nine months later, Kalu was there to welcome the retreatants back into the world and send in another group of eleven men and eleven women. These retreats have now been running back-to-back for over twenty-six years. Similar retreats are undertaken at other Kagyu practice centers. Almost a decade ago Batchelor estimated that "at any one time in Europe today up to three hundred people will be undergoing the traditional Kagyu 3-year, 3-month and 3-day meditation training."[14]

In 1967 a Soto Zen teacher, Taisen Deshimaru, arrived in Paris with the aim of establishing Zen in Europe. He was fifty-three, penniless, spoke no French, and owned no property save his notebook, monk's robe, and meditation cushion (*zafu*). Today the Taisen Deshimaru Mission has upwards of two hundred practice centers and groups internationally, ninety of which are in France; the rest are mainly in Europe but some are as far-flung as Uruguay and Cameroon. Also, the southwestern French town of Sainte Foy la Grande is the site of Plum Village, the home base of the Vietnamese Zen monk and internationally known peace activist and poet Thich Nhat Hanh.

AMERICA THE BUDDHA FULL

It was in sources like William Jones's *Asiatick Researches* that American writers Henry David Thoreau and Ralph Waldo Emerson first discovered the mind of India. Yet as Jones's own studies were confined to Hinduism, the Indian ideas that the American Transcendentalists picked up were largely Hindu. Buddhism had not yet been clearly extricated from its parent. Around 1840, Brian Hodgson, a British civil servant stationed in Nepal, happened upon a Sanskrit copy of an important Mahayana Buddhist Sutra, the *Saddharma Pundarika* ("Lotus of the True Teaching"). It found its way to Eugene Burnouf in Paris, whose French translation of it reached Emerson and Thoreau in Concord, Massachusetts. In 1844, Thoreau published an English translation of it, and (figuratively speaking) the first faint strains of "America the Buddha Full" could be heard drifting across the waters of Walden Pond.[1]

Perhaps more significant for American Buddhism than what Thoreau translated, however, was the way he lived his life—his contemplative habits. Although Emerson was first, last, and always a writer, Thoreau, in addition to writing (his journals run to fourteen volumes), experimented with silence. "Ask me for a certain number of dollars if you will," he said, "but do not ask me for my afternoons."[2] He tells us that he could sit alone for hours, listening to nature, *doing* nothing, just *being*. This wholly different kind of study deepened him, helping him grow, he said, "like corn in the night."

Thoreau was certainly not the only American contemplative, but as historian Rick Fields says, he was one of the few to live contemplatively in a Buddhist way. That is to say, he was perhaps the first American to explore the nontheistic mode of contemplation that is the distinguishing mark of Buddhism.[3] Perhaps it was a Buddhist sort of simplicity that caused Thoreau to remark that "A man is rich in proportion to the number of things he can afford to let alone," and to add that "Civilization, in the real sense of the term, consists not in the multiplication, but in the deliberate and voluntary reduction of wants."[4] And perhaps it was a Buddhist sort of gentleness that moved him to say, "I know some will have hard thoughts of me, when they hear their Christ mentioned beside my Buddha, yet I am sure that I am willing they should love their Christ more than my Buddha, for love is the main thing."[5]

One of the most potent seeds to be sown in the young American Buddha-field was the publication in 1879 of Edwin Arnold's already alluded to rhymed-verse biography of the Buddha, *The Light of Asia*. Arnold was an English poet and journalist married to an American. He had lived in India for a time, retained a lively feel for its sights and sounds, and had been deeply impressed by the Buddha's life—Asia's candidate for the greatest story ever told. Informed by the available European scholarship on Buddhism,

Arnold turned all these influences to marvelous account in a perceptive, highly sympathetic, and inspirational epic poem about the "World-honoured One." The American jurist Oliver Wendell Holmes, not known for breathlessness, wrote a twenty-six-page rave in *The International Review,* calling *The Light of Asia* "so lofty that there is nothing with which to compare it but the New Testament." He was not alone in his enthusiasm. It created a minor sensation in America, going through eighty editions and selling between a half and one million copies.

No sooner were the riches of spiritual India glimpsed by a widening American public than its visionaries began to think out loud about the possibility of a spiritual unity that lay hidden beneath religions' different exteriors, East and West. Two such were Madame Helena Blavatsky and Colonel Henry Olcott, who founded the Theosophical Society in New York in 1875, dedicating it to the study of the wisdom traditions of all humanity in an effort to discover the spiritual laws of the universe and the keys to human spiritual fulfillment. Branches of the society soon multiplied into the thousands in America, Asia, and Europe. Although the theosophists paid homage to all the great saints and sages of humankind's religious past, it was also clear that their leaders considered the Buddha a first among equals. In 1880 Blavatsky and Olcott[6] traveled to Sri Lanka and, on May 25, kneeling before a Buddhist monk, took refuge in the Triple Gem—the Buddha, the Dharma, and the Sangha. It was the first time Americans had made this formal declaration of allegiance to the Buddhist tradition.

Today the United States is the most religiously plural nation history has ever known. If there is a single moment that anticipated this remarkable fact, it was the World Parliament of Religions held in Chicago in 1893. The parliament's organizers foresaw that the increasing globalization sure to come in the twentieth century would bring with it the need for religions to enter into dialogue to reduce the intolerance that has plagued much of religious history. More than ten thousand letters of invitation were sent to religious representatives around the world, and they hit a nerve. The enthusiasm of the response surprised even the most optimistic of the organizers. Though most of the delegates to the parliament were Christians of various stripes—alone remarkable, given the prevalent interdenominational prickliness at the time—there was also a large contingent of Asian religious delegates, from Japan and India mainly, but also from Thailand, China, and Sri Lanka.

In fact, the parliament was a pivotal moment for the evolution of Buddhism in America. Zen and Theravada Buddhist traditions were well represented. Particularly sensational—alongside the dramatic presentation of Hinduism by Swami Vivekananda—were talks by Anagarika Dharmapala, a Sri Lankan Buddhist visionary and reformer who captivated audiences with his passionate conviction that the time was finally ripe for the Buddha's wise, rational, and irenic approach to life to make its claim upon the world's attention.[7] A few days after the parliament closed, Dharmapala was asked to give another talk in Chicago. As it ended and people rose to leave, someone announced that a special event was about to unfold. Charles T. Strauss, a late-twentyish New York City businessman of Jewish ancestry with a background in comparative religion and philosophy, stepped forward to take refuge in the Buddha, the Dharma, and the Sangha. Thus, on September 26, 1893, Strauss became the first person to formally join the Buddhist Sangha on American soil.[8]

A little more than a century after the World Parliament there are about three million Buddhists in America. Roughly two-thirds are Asian immigrants, or *ethnic Buddhists,* who brought their local Buddhisms with them. The remainder are American converts who fall into two classes: those who have been evangelized by Buddhist immigrants *(evangelical Buddhists)* and those who have become Buddhist on their own initiative. Most American converts belong to the latter class and, like their prototypes, Blavatsky, Olcott, and Strauss, actively sought out Buddhism on their own, usually first through reading. For want of a better term we will call them *new Buddhists,* and after a quick look at the other two groups the balance of the book will be given to them.[9]

Ethnic Buddhism. By 1852 the California gold rush had brought twenty thousand Chinese to California. By 1870 that number had more than tripled. Chinese temples, blends of Confucian, Taoist, and Mahayana Buddhist sensibilities, began to dot the California landscape. Exclusively Buddhist organizations have appeared mostly in the last fifty years. There are approximately 125 Chinese Buddhist organizations in the United States, more than half of them in California and a fifth of them in New York. One is the Hsi Lai Temple near Los Angeles, the largest Buddhist monastic complex in the West, completed in 1988 on a twenty-acre site at a cost of over $30 million. It houses a Buddhist university and a press, and serves as the world headquarters of the Buddha's Light International Organization, which has more than one hundred regional chapters. Its ecumenism has attracted Theravadins, Mahayanists, and Vajrayanins into its programs, and its membership has moved beyond the ethnic Chinese community to embrace other Asian and European Americans. Another is the Sino-American Buddhist Association, whose founder and longtime leader, the illustrious Hsuan Hua, acquired a large tract of land in California's Ukiah Valley in 1959, which he dubbed the City of Ten

Thousand Buddhas. Its 488 acres and over sixty buildings house the Dharma Realm Buddhist University and Gold Mountain[10] Monastery. In the latter, monks and nuns practice teachings drawn from all five of the main schools of Chinese Buddhism—Ch'an, Vinaya, Tien-T'ai, Tantra, and Pure Land. The Association has eight other monasteries and hermitages in the United States.

Japanese workers began to arrive in Hawaii in significant numbers in the late 1880s, and a Pure Land Buddhist temple was up and running by 1889, a year after Hawaii had been annexed by the United States. In 1898 Japanese Pure Land missionaries launched the Buddhist Mission of North America on the U.S. mainland, and in 1944 it renamed itself the Buddhist Churches of America. Headquartered in San Francisco and with over sixty churches under its wings, it is the largest and most organized of the various ethnic Buddhist enclaves.

Later-arriving Korean and Vietnamese immigrant communities also brought their local Buddhist traditions with them, which, like those of the Chinese and Japanese communities, were largely Mahayanist. The still younger Laotian, Thai, Cambodian, and Burmese communities have brought with them forms of traditional Theravada Buddhism.

Evangelical Buddhism. The prime American example of evangelical Buddhism is the Soka Gakkai ("Value Creation Society"). At the root of Soka Gakkai are the ideas of the thirteenth-century Japanese Buddhist priest and reformer Nichiren, who became convinced that the meditative way of the Buddha's Eightfold Path was no longer feasible for very many people. He held the *Lotus Sutra* to be the epitome of Buddhist wisdom and a repository of tremendous spiritual power. It is not necessary to read the Sutra, only to revere it—by repeatedly invoking the words *Namo Myoho Renge*

Kyo, "Praise to the Lotus of the Good Law." Devotees credit this practice with bringing them both spiritual and material benefits. The priest-governed Nichiren-Shoshu (Nichiren sect) prospered over the centuries and went through a period of particularly explosive growth after World War II. The Soka Gakkai, founded as a lay-governed affiliate of the Nichiren-Shoshu in 1930, experienced similar postwar growth and has become one of modern Japan's largest and most prosperous religious bodies, actively involved in its politics. It came to America in the 1960s, headquartered in Santa Monica, as the Nichiren Shoshu of America; by 1974 it boasted 258 chapters and over 200,000 members with a large contingent in Los Angeles.

We can now turn to the third American Buddhism, which—because it is the most indigenous to America and most likely to interest the readers of this book—will occupy us from here on out.

≈≈≈

ADAPTATIONS:
The New Buddhism

What is new about the New Buddhism of America? Five things: It is meditation-centered and largely a lay phenomenon. It exhibits gender parity. And it is cross-pollinating. It is socially and politically engaged.

1. 2. It is *meditation-centered* and a *lay phenomenon*. The first two items must be considered together. Down through twenty-five Asian Buddhist centuries, monks and nuns have been the tradition's vanguard, and meditation has been almost exclusively their province (and often only for an elite fraction of them). The vast majority of Buddhist laity have limited their concerns to the earning of merit—the accumulation of good karma leading to better rebirth through ethical conduct and ritual observance. The New Buddhism of America, however, has upset this traditional arrangement. First, it is largely a lay movement. And second, meditation is not the province of a relatively few specialists, but the basic practice of the many.

A recent sociological study finds that among American Buddhists, 92.4 percent ranked meditation as the single most important activity that their group carries on[1] and, the study's author says, "If there is a single characteristic that defines the new Buddhism for most of its members, it is the practice of meditation."[2] Over the first sixty years of the twentieth century, 21 Buddhist meditation centers had been founded. Between 1964 and 1975, 117 new centers were established. Between 1975 and 1984, 308 more were added. And between 1985 and 1997, 608 new meditation centers entered the picture, more than doubling the number that had been in existence until then and bringing the American total to well over 1,000. The New Buddhism of America has therefore given us something we've never seen before: a Buddhism that is predominantly *lay and meditation-centered*.

Two important corollaries follow. First, the New Buddhism has redefined the boundaries of the traditional Buddhist Sangha. At its narrowest, in the early days of the Buddha's mission, the Sangha included only monks and nuns. Soon it was understood to include all people, monastic and lay, who formally took refuge in the Buddha, the Dharma, and the Sangha, the Triple Gem. In the New Buddhism, however, there seems to be widespread if informal agreement, encouraged by widely influential Asian teachers,[3] that the Sangha includes all people who practice Buddhist meditation, whether or not they have formally taken refuge.

The second corollary is that the primarily lay makeup of the New Buddhism has loosened and in some cases virtually undone the usual lines of authority in Buddhist Asia that automatically elevated monks over laity, elders over juniors, and men over women. American Buddhist groups have been decisively influenced by the pluralist, democratic, and gender-conscious milieu in which they find themselves. Especially because of lessons learned from abuses of power that threw a number of Buddhist communities into turmoil in the 1980s,[4]

the authority of even the most charismatic teacher now appears to be open to question and discussion. Governance and
decision making in many of these groups is now in the hands
of a council or board of trustees that operates by consensus.

3. *Women and men are equals.* Although America's New
Buddhism cannot be said to have broken completely with the
legacies of gender inequality in Asian culture and Buddhist
history, Western society's trend toward gender parity is de-
parting from that legacy. In most American Buddhist groups
there are slightly more women than men. The sexes practice
together as equals and share the same roles and responsibilities in ways largely unknown in Buddhist Asia. Sociologist
James Coleman reports:

> *Although virtually all Asian and a majority of Western*
> *teachers are male, there are a growing number of*
> *women in top positions of respect and authority.*
> *Today, no one is surprised to see women leading*
> *retreats, giving dharma talks, or running major*
> *Buddhist centers. On a more theoretical level, no*
> *matter who occupies those positions of power, nearly*
> *all Western Buddhist groups recognize the full equality*
> *of the sexes and the ability of all persons of either*
> *gender to realize their true nature and attain*
> *enlightenment.*[5]

4. American Buddhism is *cross-pollinating*. The historian
Rick Fields notes that "Asian Buddhists who have not com-
municated for hundreds or thousands of years now find
themselves sitting next to one another in a new [American]
home."[6] Coleman concurs: "Never before in the long history
of Buddhism have all of its major traditions entered a new
area at the same time, and never before has there been so

much contact and exchange among these different traditions."[7] Most see this situation as an unprecedented opportunity for creative evolution; hybrids are, after all, the very
stuff of life. Proponents note that the cross-fertilization of
ideas could catalyze a revolutionary critical examination of
heretofore unexamined assumptions about sectarian superiority. Not all, however, are sanguine about the effects of the
American melting pot. Crossovers can refresh life, but biol-
ogy also teaches us that most mutations are harmful. Some
teachers (as we shall see in Chapter 18) feel that dialogue can
water down teachings and lead to ineffective syncretism.

5. It is *socially and politically engaged*. The last element of
the New Buddhism is a little different from the others. It is
not as broadly characteristic of the whole fabric as are the
other four, and to date, it remains an eddy in the larger stream.
Still, it is so important that at least one scholar has wondered
whether it will become Buddhism's fourth yana, or vehicle.
Moreover, it is not new in the sense of having no Asian
precedent. In the past eighty years Vietnam, Thailand, and
Sri Lanka have witnessed a number of important Buddhist-
based movements for social change.[8] A symbol of one of
them remains seared into the memories of many alive at the
time: the self-immolation of a Buddhist monk in protest of
the Vietnam War. And one of the inspirations for Western
Buddhist activism has been another Vietnamese Zen monk,
Thich Nhat Hanh, whose own life and political activity were
profoundly shaped by the agonies of that war. Unflaggingly
committed to the importance of the meditative life, Nhat
Hanh nevertheless remembers how the issue of sociopolitical
action arose for him:

So *many of our villages were being bombed. Along*
with my monastic brothers and sisters, I had to decide

what to do. Should we continue to practice in our
monasteries, or should we leave the meditation halls in
order to help the people who were suffering under the
bombs? We decided to do both—to go out and help
the people and to do so in mindfulness. We called it
engaged Buddhism.[9]

Like-minded New Buddhists argue that working toward
individual inner peace is not enough. What is also deeply
needed is a corresponding effort to alter social injustices in
order to lessen the suffering of humanity at large. Actually,
they say, the practice of Buddhism itself points in this direction, for the more one sees into the selfless and interdependent nature of all life—what Nhat Hanh calls "interbeing"—
the more compassion arises, not only as something to be ra-
diated from atop a meditation cushion, but as something to
be actively deployed in this suffering world. Engaged Western Buddhists have worked in hospitals, hospices, and prisons and with AIDS victims and the homeless. They have
joined in human-rights causes, antinuclear and proecological
protests, peace walks, antiwar efforts, and education for
nonviolent social change.

Much of the work of socially engaged Buddhists has been
coordinated by the Buddhist Peace Fellowship. Founded by a
few "meditating activists" on Maui in 1978 and now four
thousand members strong, it is currently headquartered in
Berkeley and is open to all. Its statement of purpose is fivefold: (1) to make clear public witness to Buddhist practice and
interdependence as a way of peace and protection for all beings; (2) to raise peace, environmental, feminist, and social
justice concerns among North American Buddhists; (3) to
bring a Buddhist perspective of nonduality to contemporary
social action and environmental movements; (4) to encourage

the practice of nonviolence based on the rich resources of traditional Buddhist and Western spiritual teachings; and (5) to
offer avenues of dialogue and exchange among the diverse
North American and world sanghas.[10]

One striking example of engaged Buddhism is the San
Francisco/Hartford Street Zen Center's Maitri Hospice for
AIDS patients founded in the late 1980s. Another is the work
of Joanna Macy, a brilliant Buddhist scholar and writer informed by both Theravadin and Tibetan traditions, who
opted out of a promising academic career to work more directly for peace and political change. Over the last twenty-
five years Macy has traveled the globe to offer countless
antinuclear and proecological workshops inspiring people to
overcome their sense of helplessness and work nonviolently
for positive change. Macy's Buddhist activism has also been
shaped by her fieldwork with Sarvodaya, a grass-roots
economic-development movement in the village communities
of Sri Lanka.[11]

Yet another example of engaged Buddhism is the Zen
Community of New York, led by Roshi Bernard Tetsugen
Glassman. It has opened a bakery that has trained hundreds
of chronically homeless persons and provided many people
with jobs. It also built the Greyston Family Inn, which se-
cures permanent housing and provides support services for
once homeless families. As an alternative to *sesshin,* the
week-long meditation intensives common in the Zen tradition, Glassman introduced the "street retreat." Participants
are required to raise $1,000 for each day of the retreat, do-
nating the funds to homeless-aid organizations. For five days
they practice—walking the streets of New York without a
dime for food, a change of clothes, or a place to sleep, themselves panhandling, supping in soup kitchens, and just sur-
viving. Their twice-daily practice of zazen (Zen sitting

meditation) during this period underscores an enduring lesson in both suffering and compassion.

The offering of social services and the struggle for social change are undoubtedly critical aspects of engaged Buddhism, but, as Thich Nhat Hanh himself points out, so are simple acts of kindness in the midst of one's daily routine. Engaged Buddhism includes a broad range of possibilities, all of which are ways of applying the Buddha's teachings, living the life he modeled, and walking the talk. It is not surprising that both Thich Nhat Hanh and the Dalai Lama, men who have had to deal with enormous human suffering for the greater part of their lives, would be among the leading contemporary proponents of engaged Buddhism. What is at least a little surprising, perhaps, is that, despite the agonies of which they are all too aware, neither ever seems to forget the Buddha's counsel that truly effective action flows only from wholesome and discerning states of mind. Either man could have written what the Dalai Lama in fact wrote in his Nobel Peace Prize acceptance speech in 1989: "Inner peace is the key. In that state of mind you can deal with situations with calmness and reason."[12]

AMERICA STARTS MEDITATING I:
The Ways of Zen

In 1997, 40 percent of the more than one thousand Buddhist meditation-oriented groups in North America were affiliated with the Zen tradition. Zen is largest single Buddhist presence in America partly because it is the oldest. Among the participants in 1893 World Parliament of Religions in Chicago was a Rinzai Zen priest named Soyen Shaku. When he returned to America in 1905 to teach Zen in San Francisco, it was for only a number of months. But three of his students came to America and stayed much longer.

One of them, Sokei-an, founded the Buddhist Society of America (later renamed the First Zen Institute of America) in New York in 1931. It was not a particularly auspicious beginning. The society had four members at first; four years later there were fifteen; seven years later, thirty. Sokei-an likened the difficulty of growing Buddhism in America to waiting for a lotus to take root while holding it against a rock. Indeed, it would be 1959 before the first major Zen practice community opened in America. What Sokei-an could not have suspected, however, was that by 1975 there would be a hundred of them.

Nyogen Sensaki came to America in 1905 as Soyen's personal attendant. Before Soyen departed, he told Sensaki not to teach Zen for seventeen years. Sensaki worked as a houseboy, a farmer, and a hotel operator during those years, likening himself to a mushroom—"without a very deep root, no branches, no flowers, and probably no seeds." But in 1922, his obedient silence over, he began to teach. Until his death in 1958 he gathered Zen groups in San Francisco and Los Angeles in a series of rented rooms he called the "floating zendo" (a *zendo* is a Zen meditation hall). With meditators seated in chairs, he taught zazen to Japanese on one night and English-speakers on another. Among the latter was Robert Aitken, later to become one of the first generation of American Zen masters. Once, in describing himself as much as he was describing his ideal, Sensaki wrote: "A Buddhist monk is celibate and he leads the simplest life possible. He never charges. He accepts used clothes or old shoes and wears them. Any excess food or money he gives away. He sleeps quietly without worries, having none in his possession." In 1945, temporarily homeless after his release from an American internment camp for Japanese, Sensaki composed the following lines on the anniversary of his teacher's death:

For forty years, I have not seen
My teacher, Soyen Shaku, in person.
I have carried his Zen in my empty fist,
Wandering ever since in this strange land.
The cold rain purifies everything on earth.
In the great city of Los Angeles, today
I open my fist and spread the fingers
At the street corner in the evening rush hour.[1]

The third student Soyen Shaku sent to America, Daisetz Teitaro (D. T.) Suzuki, cut another pattern altogether. Born in 1870, Suzuki had taught himself English from books while still a boy. When Soyen Shaku needed his World Parliament talk rendered into English before he sailed for the States, he called on Suzuki, who was then one of his novice monks. While in America, Soyen made a number of friends, one of whom was Paul Carus, a writer and editor with a deep interest in Buddhism. Upon Soyen's recommendation, Carus invited Suzuki to work with him at Open Court Publishing Company in La Salle, Illinois. Suzuki arrived in 1897 and stayed in the West, mostly in America, for fourteen years. In 1911, now married to an American, Suzuki returned to Japan to resume his Zen studies under Soyen. When Soyen died in 1919, Suzuki left the monastery and became a professor, teaching philosophy of religion at the University of Kyoto. By 1927, he had already written a couple of books on Mahayana Buddhism[2] in English, but in that year he produced *Essays in Zen Buddhism: First Series,* the fountainhead of what was to be a prodigious outpouring of Zen whose wide readership would make Professor Suzuki (who died in 1966 at the age of ninety-six, still working at his desk) the twentieth-century's single greatest influence in the coming of Zen Buddhism to America.

In 1936, Suzuki visited England to attend the World Congress of Faiths. There he met a twenty-one-year-old Englishman who had declared himself a Buddhist at the age of fifteen, and whose own later writings were to contribute perhaps only a little less greatly than Suzuki's to the evolution of American Zen. His name was Alan Watts. After World War II, Suzuki came to America for a series of lectures at Columbia University. Among those in attendance were the journalist Philip Kapleau, a future American Zen master, the psychoanalyst Erich Fromm, and the musician John Cage, each of whom would play a significant role in the Zenning of America. Beat writers Jack Kerouac and Allen Ginsberg missed Suzuki at Columbia, but caught up with him at the New York Public Library. The young Gary Snyder (whose Zen poetics would later reach huge audiences) had read a bit about Buddhism while still an undergraduate at Reed College, but it wasn't until he was passing through San Francisco on his way to graduate school in Indiana that he bought a copy of Suzuki's *Essays in Zen Buddhism* and tossed it into his backpack. When Kerouac and Ginsberg finally met Snyder in San Francisco in 1955, their shared interest in Zen and Suzuki was a key factor in their chemistry. About the same time, Snyder, attending a Buddhist study group in Berkeley, bumped into an articulate, fortyish Englishman who, after a stint as the Episcopalian chaplain at Northwestern University, had now settled in the Bay Area. It was the aforementioned Alan Watts.

Such was at least a small subset of the cast of characters behind the initial phase of American Zen's 1960s explosion, a phase in which a primarily literary Zen—Zen ideas, Zen style, but as yet not much Zen meditation—threaded its way into the American tapestry. Gary Snyder, an exception, had been sitting zazen atop his rolled-up sleeping bag for years, but Watts, for example, was too extroverted to still his mind and gave every impression of avoiding meditation cushions like the plague. About these early days, Allen Ginsberg would write: "Nobody knew much about zazen. It was a great tragedy. If somebody had just taught us how to sit, straighten the spine, follow the breath, it would have been a great discovery."[3]

Enter a second Suzuki. Shunryu Suzuki was a Soto Zen priest who arrived in the San Francisco area from Tokyo in the spring of 1959 to head the Soto Zen Mission temple in San Francisco's Japan town. Since its founding in 1934, it had been an ethnic Buddhist enclave pretty much circumscribed by the Japanese community. But in the years immediately prior to Suzuki's arrival a few Westerners began attending its services and sitting zazen. Under Suzuki Roshi the numbers increased. To Westerners who sought admission, Suzuki Roshi would give a standard reply: "I sit at five-thirty in the morning. You are welcome to join me."

If the Beats and other Zen aficionados of the late 1950s had found that—again, with the exception of Gary Snyder— sitting dampened their flamboyant style, the youth of the 1960s did not. They were dying to meditate. With Shunryu Suzuki, supply met demand. Suzuki Roshi's emphasis on the centrality of meditation practice as well as his specific instructions on how to go about it were ultimately derived from Dogen, the thirteenth-century patriarch of the Soto Zen school, considered by some to be one of the finest intellects Japan has produced. Said Dogen (who died while sitting in zazen):

> In pursuit of the Way, the prime essential is zazen. Just to pass time in sitting straight, without any thought of acquisition, without any sense of achieving enlightenment—this is the way of the patriarchs. It is true that our predecessors recommended both the koan and sitting, but it was the sitting they particularly insisted upon. Truly the merit lies in the sitting. Simply to go on sitting is the method by which the Way is made an intimate part of our lives. Thus the attainment of the Way becomes truly attainment through the body. That is why I put exclusive emphasis on the sitting.[4]

The eagerness with which many of Suzuki Roshi's students dove into meditation doubtless fed upon their vivid anticipation of the end to which meditation supposedly led—that incomparable, ineffable, burst of insight, called satori or kensho, that signaled Zen enlightenment. Satori was, after all, a constant theme in the writings of Zen scholar D. T. Suzuki, by then *de rigueur* for Zen students. But the second Suzuki had a surprise in store. In sharp contrast to the first Suzuki, Shunryu Suzuki hardly mentioned satori. In fact, while insisting on sitting, he taught that there was nothing to achieve by it. "To take this posture is itself to have the right state of mind," he said, probably much to the consternation of those who interpreted the acute discomfort of their initial zazen as enlightenment's polar opposite. Yet Suzuki Roshi quietly insisted: Sitting does not achieve enlightenment; it participates in an enlightenment that is already there. As Dogen himself had taught, practice and enlightenment are not two:

> *In Buddhism, practice and enlightenment are one and the same. Since practice has its basis in enlightenment, the practice of even the beginner contains the whole of original enlightenment. Since enlightenment is already contained in the exercise, there is no end to enlightenment, and since it is the exercise of enlightenment, it has no beginning.*[5]
>
> *To practice the Way single-heartedly is, in itself, enlightenment. There is no gap between practice and enlightenment or zazen and daily life.*[6]

It is not hard to sense the serene delight Suzuki Roshi took in helping his students see that meditation was not about having some experiences and then being finished, but about living in the world in a different way, about a way of life that is endlessly refinable. "When a fish swims in the water there is no end," he quoted Dogen as saying, while adding,

> *It is very interesting that there is no end. Because there is no end to our practice it is good. Don't you think so? Usually you expect our practice to be effective enough to put an end to our hard practice. If I say just practice hard for two years, then you will be interested in our practice. If I say you have to practice your whole lifetime, then you will be disappointed. You will say, "Oh, Zen is not for me." But if you understand that the reasons you are interested in this practice is because our practice is endless, that is true understanding. That is why I am interested in Buddhism. There is no end.*[7]

Suzuki Roshi's teachings remained largely within the walls of the San Francisco Zen Center, which he and his community founded in the early 1960s, until the appearance in 1970 of a little book of his talks. Enormously popular, *Zen Mind, Beginner's Mind* continues to exert an influence on American Buddhism thirty years after its publication.

The San Francisco Zen Center has continued to be a major American Buddhist hub long after Suzuki's death in 1971, producing a new generation of American Zen Buddhist teachers. A similar story unfolded in southern California, where Taizan Maezumi Roshi founded the Zen Center of Los Angeles in 1968. Maezumi's teachers had been Harada Roshi and Yasutani Roshi, who attempted to blend the ideas and practices of the Soto and Rinzai lineages. Before his death in 1995, Maezumi left twelve dharma-heirs,[8] a number of whom have founded their own thriving Zen communities in other parts of the United States. Two outstanding examples of these are the above mentioned Bernard Tetsugen Glassman's Zen Community of New York, for example, and John Daido Loori's Zen Mountain Monastery[9] in Mt. Tremper, New York.

Studying in the same lineage as Maezumi were Americans Philip Kapleau (b. 1912) and Robert Aitken (b. 1917). Kapleau had gotten interested in Zen while a court reporter for the International War Crimes Tribunal in Tokyo in 1948. After thirteen years of training in Japan, he founded the Zen Meditation Center of Rochester in New York in 1966, a year after he had published *The Three Pillars of Zen,* the first major book on Zen written by a Westerner. Also a proven classic, *Pillars* stands in sharp contrast to Shunryu Suzuki's *Zen Mind.* The latter is a wispy, tantalizingly indirect series of reflections in no particular order. The former is a veritable storehouse of information, magisterial in its organization and lucidly direct in its style. Kapleau Roshi, like the Rinzai lineage in which he was partly formed, was skeptical about what he saw as the downplaying of the satori/enlightenment experience by Zennists under Dogen's influence. In *Pillars,* satori is very much in clear view, and one of the book's major sections includes personal accounts by men and women who have experienced it. But both books and both roshis were in deep agreement about at least one thing: that the heart of Buddhism, the heart of Zen, is the practice of zazen. Kapleau's work was probably the first to include detailed instructions on how to sit. Many who had never met a Zen teacher or lived near a Zen group took their first steps on the meditative path through his book.

Robert Aitken discovered Zen while a prisoner of the Japanese during World War II. Stimulated by repeated readings of R. H. Blyth's *Zen and English Literature,* a book loaned him by a guard, Aitken was amazed to run into Blyth as a fellow prisoner. Their conversations over a period of fourteen months launched Aitken's postwar interest in zazen. He met one of his beloved teachers, Nyogen Sensaki, during a brief stay in Berkeley and experienced the first of many sesshin (Zen practice intensives) in Japan in 1950. In 1959, having settled in Hawaii, Aitken and his wife, Anne Hopkins, began a zazen group called the Diamond Sangha and established the Koko-an Zendo in Honolulu. Ten years later, after an extended period of further study under Yasutani Roshi in Japan, Aitken, along with his wife, moved to Maui and founded the Maui Zendo. There, in 1978, Aitken and others founded the Buddhist Peace Fellowship (see Chapter 15) "to bring a Buddhist perspective to the peace movement, and to bring the peace movement to the Buddhist community." A Zen master, activist, and authority on the application of Buddhist ethics in the modern world, Aitken Roshi has many students who have gone on to form their own Zen communities.[10]

Walter Nowick has been called the quietest Zen master in America. Few know that he was the first American to receive full transmission in an orthodox line of Rinzai Zen. A Juilliard-trained musician who once studied at the First Zen Institute of New York, Nowick worked for years in Japan under Zuigan Goto Roshi before becoming one of the latter's dharma-heirs. Nowick Roshi's teaching and rural Maine zendo are charmingly—though anonymously—recalled in J. van de Wetering's perceptive Zen memoir, *A Glimpse of Nothingness*.

Women have played an important role in the flowering of American Zen as students and teachers. Blanche Hartman and Linda Ruth Cutts, students of Suzuki Roshi, have served as abbesses of the San Francisco Zen Center. Yvonne Rand, a practitioner since she met Suzuki Roshi in 1966, has also been a central figure in Zen Center's rise to prominence. She brings a pro-choice, anti-abortion Buddhist perspective to reproductive issues by defending a woman's right to choose while teaching that abortion's moral gravity makes it at best an option of last resort. Jan Chozen Bays and Joko Beck received dharma transmission from Maezumi Roshi in Los Angeles; the former now teaches in Oregon, the latter heads the San Diego Zen Center. Jiyu Kennet, an Englishwoman trained in Japan, received dharma transmission in a Soto lineage and founded Shasta Abbey in northern California. Maurine Myoon Stuart, the late head of the Cambridge Buddhist Association had studied Rinzai Zen under Eido Tai Shimano of the Zen Studies Society in New York and was herself made roshi by Eido's teacher, Soen Nakagawa Roshi in 1982. Roko Sherry Chayat, also a disciple of Eido Roshi, heads the Zen Center of Syracuse, New York. Toni Packer was one of Philip Kapleau Roshi's leading students and presumed successor until her skepticism about the relevance of traditional Japanese forms in the practice of American Zen led to a break. In 1981 Packer and her students founded the Genesee Valley Zen Center, now the Springwater Center in Springwater, New York. While Zen tradition often speaks about the cultivation of no-mind, Packer's approach to meditative inquiry, influenced by the iconoclastic teachings of J. Krishnamurti, might be called no-Zen or even no-Buddhism. Packer appears to be stretching to new limits the form-busting originality celebrated in the annals of the Zen masters.

Not all of Zen's American presence has its roots in Japan. The charismatic Korean Zen master Soen Sa Nim (b. 1927) arrived in America in 1972 and, along with his students, has founded Zen Centers in Providence, Cambridge, New Haven, New York, Los Angeles, and Berkeley. Vietnamese Zen has been represented in part by the already mentioned Thich

Nhat Hanh (b. 1926; see Chapter 15), who heads an international network of sanghas known as the Community for Mindful Living; and also by Thich Thien-an (1926–80), honored as the first Vietnamese American patriarch of Buddhism. Thien-an came to the United States in 1966 as a visiting professor at UCLA. By 1970 the meditators he instructed on the side had become the International Buddhist Meditation Center in Los Angeles. Three years later, he founded the University of Oriental Studies. Though trained in the Lin Chi (Rinzai) tradition of Ch'an (Zen) Buddhism, Thien-an's style was ecumenical, attracting Theravadin, Tibetan, Korean, and Japanese Shingon teachers to the university.

Even a slightly less sketchy review of the American Zen wave would have required far more space than we have available. Fortunately we can refer interested readers to excellent resources.[11]

AMERICA STARTS MEDITATING II:
Tibetan Buddhism in Exile

Zen has been in America since the beginning of the twentieth century. Tibetan Buddhism did not appear until a half century later, but it burgeoned in the 1970s and exploded in the following two decades. Between 1987 and 1997, Vajrayana groups in North America doubled from 180 to 360 and today account for a little more than a third of Buddhist meditation-oriented groups.

It all began modestly enough when, in 1955, a Mongolian monk and scholar, Geshe Wangyal, moved to Freehold Acres, New Jersey, where a small community of Mongolians, displaced by World War II, had already settled. There were a few Tibetan temples in the area, serving pretty much as community centers, but Geshe Wangyal (*geshe* is the Vajrayana equivalent of Ph.D., a title awarded only after seventeen years of rigorous study) envisioned something more. Geshela, as his students affectionately called him, founded the Lamaist Buddhist Monastery of America, the first Tibetan monastery in the United States, now renamed the Tibetan Buddhist

Learning Center. He was soon joined by two young Tibetan *tulkus* (tulku is a title given to one believed to be the reincarnation of a former illustrious Tibetan lama) and another Tibetan geshe, Geshe Sopa, who was later to become a key figure in the seminal Buddhist studies program at University of Wisconsin in Madison.

Having gotten word of Geshe Wangyal's presence, some Harvard students drove down to meet him and to ask for teachings. He agreed to teach them—yes, meditation and the Vajrayana path that were uppermost in their minds, but first and foremost, as befitted a geshe's scholarly bent, the Tibetan language. The tuition he required would be that they teach English to the young tulkus.

One of the those students was Robert Thurman. Thurman became fluent in Tibetan and so relished the Tibetan teachings that by 1964 he pressed Geshe-la to ordain him. Geshe refused, but, seeing that Thurman was not deterred, suggested that he go to India and make the same request of the Dalai Lama. Geshe Wangyal accompanied Thurman to India and introduced him to His Holiness, who encouraged Thurman to continue his studies there. Within the year, Thurman became the first American to be ordained as a Tibetan Buddhist monk.

When Thurman returned to the United States, he found his monkhood ill-fitting and disrobed, but his love of the Dharma was undiminished. Encouraged by Geshe Wangyal, he continued his Tibetan studies and completed a doctorate at Harvard. Now the Jey Tsong Khapa Professor of Indo-Tibetan Buddhist Studies at Columbia University, the ebullient Thurman is one of the most respected interpreters, translators, and shapers of American Vajrayana. Despite his own earlier decision to become a Buddhist layman, Thurman is a staunch advocate for the establishment of the monastic traditions of Asia in the West as the necessary substructure for a successful, long-term transmission of the Dharma in the Western world. In New York in 1987, he cofounded Tibet House, a cultural center and art gallery that educates the public about Tibet's spiritual and cultural wealth. In 1997, *Time* named him one of the twenty-five most influential Americans. In his recent *Inner Revolution* Thurman goes further than any thinker we know in envisioning not only an American Buddhism, but a Buddhist America. It is time, he suggests, for the great American experiment in freedom, born in the "hot" revolution of 1776 and nurtured by the "outer modernity" of the European Enlightenment, to grow to maturity in the "cool" revolution of the Path and the "inner modernity" of Buddhist enlightenment.

Also in that early group of Geshe-la's students was Jeffrey Hopkins, now Professor of Indo-Tibetan Studies at the University of Virginia, an important American center for the study of the Tibetan language and for the cultural preservation of Tibet through the collection, preservation, translation, and dissemination of texts.[1] Thurman, Hopkins, and other students of Geshe Wangyal have also founded the American Institute of Buddhist Studies, aiming to bridge academia and Buddhist philosophical teachings.

The occupation of Tibet by mainland Chinese forces that began in 1950 continues to this day. In 1959, five years before Thurman met him, the Dalai Lama escaped the Chinese incursion into Tibet and crossed the Himalayas into India. One hundred thousand other Tibetans joined the Dalai Lama in exile before the Chinese were able to seal the border. Twenty thousand of these were lamas who reconstructed their monasteries on Indian soil. The Tibetan diaspora reached American shores in force in the 1970s and today there are about ten thousand Tibetan exiles in America. Lamas from all four of the major Vajrayana orders—the Gelug, the Sakya, the Nyingma, and the Kagyu—teach in the United States.

The Gelugs and the Sakyas are generally associated with a Buddhist training regime that emphasizes language mastery and philosophical study, as the young Thurman and Hopkins discovered. Geshe Wangyal was a Gelug, and the Tibetan Buddhist Learning Center he founded is one of the premier Gelug communities of the West,[2] combining a program in text translation with traditional Vajrayana training. On several occasions since 1979, the center has played host to a fellow Gelug, the Dalai Lama, whose visibility as a world religious leader has become second only to that of the pope. His remarkable charisma, widely read books, 1989 Nobel Peace Prize, and Gandhi-like commitment to nonviolence in the struggle for Tibetan autonomy have put a warm and attractive face on Buddhism for many Americans and exerted a profound influence on American Buddhism's evolution. Other important Gelug centers in the United States are Lama Zopa's Osel Shen Phen Ling in Missoula, Montana, the Namgyal monastery's North American branch in Ithaca, New York, and Geshe Sopa's above mentioned Tibetan study center in Madison, Wisconsin.

A final Gelug contribution should be mentioned: the multiphonic chanting of its Gyuto monks (see Chapter 10). In introducing Tibet to the American public, the mesmerizing performances of these monks (staged by Mickey Hart of the Grateful Dead) may have been exceeded only by those of His Holiness and Robert Thurman.

The root of Sakya presence in America was the renowned scholar Deshung Rinpoche, who arrived at the University of Washington in 1960 to participate in a research project sponsored by the Inner Asia program. In 1974, along with Jigdal Dagchen Sakya Rinpoche, he founded a small dharma center, which has now evolved into the Sakya Monastery of Tibetan Buddhism (glimpsed in the 1993 feature film *Little Buddha*). The monastery takes an integrative, nonsectarian approach to Tibetan teachings called *rimed* (pronounced ree-may) and has hosted lamas from all four traditions. Its rigorous training program is supported by an extensive library and coexists with an active calendar of public events. Deshung Rinpoche's 1980 visit to Cambridge, Massachusetts, occasioned the founding of Sakya Sheidrup Ling, or Institute for Buddhist Studies and Meditation. In 1971, Lama Thartse Kunga, a Sakya brought West by Geshe Wangyal nine years earlier, founded the Ewam Choden Tibetan Buddhist Center in Kensington, California, which offers Tibetan language training and meditation instruction. A similar center for Tibetan studies and meditation, Sakya Phunstok Ling, was founded in Silver Springs, Maryland, in 1986.

The other two Tibetan orders, the Nyingma and the Kagyu, put relatively less emphasis on formal learning and more emphasis on meditation and the performance of Tantric ritual. The Nyingma lineage founder, the legendary tenth-century saint Padmasambhava, is said to have prophesied that "when the iron bird flies, and horses run on wheels, / The Tibetan people will be scattered like ants across the World, / And the Dharma will come to the land of the Red Man." The first major Nyingma pioneer in America was Tarthang Tulku, who established the Nyingma Meditation Center in Berkeley in 1969. One of the characteristics of the new American Buddhism, the meshing of serious Buddhist practice with lay American life-in-the-world, has certainly been in evidence here. As is common in Tibetan training, beginning students were asked to perform one hundred thousand full-body prostrations (over a period of months). Rick Fields recalls that the "sheer physical labor, the barbaric reality of it, soon took care of any mystical daydreams about great yogic powers. Prostrations brought one down to earth, quite literally."[3] But Tarthang also stressed the centrality of meditation: "Meditation is the essence of the Buddha Dharma. Discussing it and thinking about it won't help. We must practice. In the beginning, meditation seems separate from us but eventually it becomes our own nature."[4]

The Nyingma Center's Dharma Publishing has undertaken a Herculean labor of love in producing the *Kangyur* (the words of the Buddha) and *Tengyur* (commentarial traditions), which comprise the 128-volume canonical scripture of Tibetan Buddhism, in a gilded, richly illustrated format. Each atlas-sized, ten-pound, handbound volume is about four hundred pages long and printed on acid-free paper to ensure its survival for centuries. Since 1975 in Sonoma County, California, the Nyingma community has been building a nine-hundred-acre "temple city" surrounded by a natural preserve and animal sanctuary. Called Odiyan, it has as its purpose the support of both long- and short-term intensive Buddhist practice. Tarthang himself seems to have disappeared from public view and is said to be devoting his time and energy to working with his advanced students.

Another Nyingma teacher, Sogyal Rinpoche, has written one of the most widely read books on Tibetan Buddhism in recent years, *The Tibetan Book of Living and Dying*. As a young boy he was identified as a reincarnation of Terton Sogyal, a master of the *dzogchen*[5] practice he continues to espouse. Sogyal's teachers were Dilgo Khentsye Rinpoche and the late head of the Nyingma lineage, Dudjom Rinpoche. In 1981 Sogyal and his students founded the Rigpa Fellowship, an international network of Buddhist centers with a major hub in Santa Cruz, California.

The Kagyus were the last of the Tibetan orders to enter

America, but have been very busy. Kagyu-linked American centers and groups abound. The person most responsible for Kagyu efflorescence in the West and America has been Chogyam Trungpa Rinpoche. Trungpa had fled across the Indian border about the same time as the Dalai Lama. He studied in India for a time, completed a Western education at Oxford University, and arrived in North America in 1970. Immensely enterprising and stirred by the ancient Tibetan myth of Shambhala, a hidden valley where an enlightened society lives in peaceful and joyous appreciation of all life, Trungpa and his devoted American students pooled their prodigious energies toward realizing it. In addition to meditation centers, Trungpa and his students have built a preschool, an elementary school, an accredited institution of higher learning that offers both undergraduate and graduate degrees (the Naropa Institute, in Boulder, Colorado), a credit union, bookstores, and numerous for-profit businesses. Trungpa-begotten dharma centers and study groups in America and beyond now exist under the umbrella of Shambhala International. They include the Karma Choling meditation and retreat center in Barnet, Vermont, the Rocky Mountain Shambhala Center in northern Colorado, and a solitary retreat center in southern Colorado. Trungpa died in 1987. His eldest son, Mipham Rinpoche, now acts as head of the organization.

Tucked within Tibetan Buddhism's four main schools is a modern ecumenical, transsectarian movement called rimed. Like many Kagyu teachers before him, Trungpa was comfortable with new integrations and was a proponent of rimed. Another important, innovative voice in American Vajrayana, Lama Surya Das (Jeffrey Miller), also aligns himself with rimed. He encountered Tibetan Buddhism while traveling in India in the 1960s and 1970s and, in the following decade, undertook two traditional three-year, three-month, and three-day retreats under the direction of the Nyingma masters Dilgo Khyentse Rinpoche and Dudjom Rinpoche. His training in dzogchen[6] led to his founding in 1991 of the Dzogchen Foundation, in Cambridge, Massachusetts. Believing that Westerners are often fairly sophisticated psychologically, Surya Das has raised eyebrows by forthrightly offering his students a teaching traditionally considered to be at the end of a long road. His efforts, including a popular book called *Awakening the Buddha Within: Tibetan Wisdom for the Western World,* have made him one of the most influential teachers on the American scene. Surya Das's interests in dialogue and integration reach beyond the Tibetan milieu to American and European teachers of vipassana and Zen. He has been instrumental in organizing the Western Buddhist Teachers Network to help respond to the challenges all lineages face in introducing the Dharma to the West.

The first American to be named a geshe is Michael Roach, who received that distinction after twenty-two years of study in India and America. A fully ordained Gelug monk, Roach is also a scholar of Sanskrit, Tibetan, and Russian and a popular Buddhist teacher and lecturer on the American circuit since 1981. He is perhaps best known as the director of the Asian Classics Input Project (ACIP). Under the guidance of Geshe Roach, Tibet's above mentioned 128-volume canon of Kangyur and Tengyur, along with noncanonical commentaries and dictionaries related to them (in all, the cultural equivalent of a DNA genome), is being transferred to searchable CD-ROMs available at nominal cost to scholars and other interested parties. Gelek Rinpoche, a Tibetan tulku and teacher, celebrates the event in a poem entitled "In Praise of the ACIP CD-ROM: Woodblock to Laser":

A hundred thousand
Mirrors of the disk
Hold the great classics
Of authors
Beyond counting;
No longer
Do we need
To wander aimlessly
In the pages of catalogs
Beyond counting.

With a single push
Of our finger
On a button
We pull up shining gems
Of citations,
Of text and commentary,
Whatever we seek;
This is something
Fantastic,
Beyond dreams.[7]

As traditional Buddhist practice continues to mesh with American lay life, Vajrayana communities are experimenting with new uses of traditional terminology. *Lama,* formerly a term used only to identify an ordained Tibetan monk, is now used by students of Kalu Rinpoche, a renowned Kagyu yogi and teacher, to describe anyone who completes a three-year, three-month, and three-day solitary retreat.[8] Many Europeans and Americans have now completed this traditional training thanks to the efforts of Kalu, who first made it available to thirteen Westerners (seven men, six women) in France in 1976. A growing number of senior students in Trungpa's line have been named *acharyas* ("teachers") and authorized to start meditation groups of their own.

Characteristic of America's New Buddhism, a number of acharyas are women. One, Pema Chodron, is a prolific author and the operational head of a monastic center called Gampo Abbey, in Cape Breton, Nova Scotia, established in 1985 explicitly for the training of Westerners. Another Kagyu, Karma Lekshe Tsomo, has combined an academic career specializing in Buddhist women's issues with her life as a fully ordained nun. In 1987 she helped found Sakyaditta ("Daughters of the Buddha"), an international association of Buddhist women working toward improvement of education in the Dharma, improved facilities for study and practice, and the establishment of communities of fully ordained women where they do not currently exist. A fully ordained Gelug nun, Thubten Chodron is a senior teacher with the Foundation for the Preservation of the Mahayana, an international network of practice centers founded in the 1970s by Lamas Thubten Yeshe and Thubten Zopa Rinpoche.

Another woman whose story suggests the radically new ground American Vajrayana is breaking is Catherine Burroughs. As a young, married, native New Yorker, Burroughs moved to rural North Carolina and was so isolated that she had little contact with anyone other than her husband. Apparently spontaneously and without prior instruction, she found herself practicing a form of Tibetan meditative body awareness. Fifteen years later, in an airport in Washington, D.C., Burroughs met Penor Rinpoche, head of a Nyingma lineage. Something clicked. Burroughs then visited Rinpoche in India, where he informed her that she was an incarnation of a great woman yogi and saint of the seventeenth century. Accordingly, in 1988, Penor Rinpoche officially installed Burroughs, now called Jetsunma, as the head of a large Vajrayana practice and study center in Poolesville, Maryland, where she lives with her husband and two children.

❧❧❧

AMERICAN STARTS MEDITATING III:
The Vipassana Movement

The American vipassana movement is growing rapidly. In the ten years between 1988 and 1998, the number of vipassana organizations more than doubled from 72 to 152 and now account for about 15 percent of Buddhist meditation groups and centers in North America.

Of the three American streams of meditative Buddhism discussed in this book, vipassana has been the most liberally adaptive, the one least tied to the monastic traditions in its Southeast Asian home. This is surprising because for over two thousand years vipassana has been preserved and transmitted by the Theravada school, the most conservative branch of Buddhism. What brought this to pass?

The foundations of American Zen and American Vajrayana were laid largely by immigrant Japanese and Tibetan monks who taught within the traditional ceremonial forms of their home cultures. The spiritual authority of the master, the centrality of the master-student relationship, the importance of rituals such as chanting, bowing, prostration, and dress—American disciples in these traditions received such practices as integral parts of the teaching. The foundations of American vipassana, however, were built by American lay-people who studied it in India, Burma, and Thailand and then returned to the West to make it available to others. Just as important, the vipassana these Americans found in South-east Asia had *already* been significantly distanced from its traditional Theravada monastic contexts. Burmese monks like Ledi Sayadaw and Mahasi Sayadaw (1904–82), to whose lineages a great deal of American vipassana can be traced, were instrumental in taking Buddhist meditation out of its monastic confines and offering it to a wider public.[1] They built "meditation centers" (a notion unknown before the twentieth century) expressly for this purpose. Since 1949, Mahasi Sayadaw's centers in Yangon (Rangoon) and other locations in Myanmar (Burma) have taught vipassana to up-wards of seven hundred thousand people, including many Westerners. In the 1950s, Ledi Sayadaw's dharma-grandson, the late Burmese layman U Ba Khin (1899–1971), opened a meditation center in Yangon that has been similarly influen-tial.

The American vipassana movement has therefore been more laicized, and even more exclusively focused on the practice of meditation, than its Zen and Vajrayana cousins. Ritual and ceremony are relatively absent. American vipas-sana teachers tend to present themselves as *kalyana-mitta*, "spiritual friends," authorized to teach because they've med-itated for more years than those they are teaching. The guru devotion that is central to the Tibetan path finds little echo in the vipassana movement. Nor does the powerful figure of the roshi, who certifies a Zen student's progress in awakening.

The trend toward gender parity that is characteristic of American Buddhism as a whole also holds for the vipassana movement. Females often outnumber males at meditation retreats. There are many women vipassana teachers, and women are also well represented on the boards and trusts that govern American vipassana organizations.

The foregoing applies to the vipassana movement as a whole. The rest of its picture falls roughly into halves—"roughly," because there are aspects of American vipassana that fall outside these convenient halves and because we must allow a few names to stand for many. On the one side stand Joseph Goldstein, Jack Kornfield, and Sharon Salzberg, who pioneered an affiliation of Insight Meditation teachers and retreat leaders that now includes at least sixty others,[2] for example, Ruth Denison, who trained in Burma and founded the Desert Vipassana Center in southern California, and Sylvia Boorstein, a popular Buddhist author.[3] On the other stands the Indian teacher Satya Narayan (S. N.) Goenka, who, along with over a score of senior American assistant teachers, has disseminated his own type of vipassana practice that stems from the aforementioned U Ba Khin.

The two sides are not strangers to each other. Joseph Goldstein and Sharon Salzberg, for example, were among Goenka's first Western students, and Goenka regards them warmly as his dharma-children. The sides share a deep appreciation for the Buddha's Way and an equally deep commitment to sowing its seeds wherever these are welcome. Indeed, the shared values that lie behind their divergence make their relationship complementary rather than oppositional, a *yin* and *yang* of American vipassana. Still, there is a significant difference between them. Aside from relatively minor differences in meditation technique, retreat format, and organizational structure, it comes down (as we shall see) to a disagreement over degree of assimilation.

Jack Kornfield (b. 1945) and Joseph Goldstein (b. 1944)—

Jack and Joseph as they are affectionately called—came in contact with vipassana in the 1960s while working in the Peace Corps in Southeast Asia, Jack in Thailand, Joseph in India. Jack's teacher was the Thai forest monk Ajahn Chah, but he also studied later with the Burmese monk Mahasi Sayadaw. Joseph's first teacher was Anagarika Munindra, a disciple of Mahasi Sayadaw, but later he studied with Mahasi himself and, in the 1980s, with U Pandita, a monk in Mahasi's lineage. Joseph met Sharon in one of the many courses she took with Goenka, but Sharon later studied intensively with U Pandita. The point of this genealogy is simply to discern a common ancestor. As the vipassana teacher and scholar Gil Fronsdal says, this branch of the American vipassana movement "is based on the systematization of vipassana meditation developed and propagated by the Burmese monk and meditation teacher Mahasi Sayadaw."[4]

Jack and Joseph did not meet until 1974, when both were invited to teach meditation during summer session at the Naropa Institute in Boulder, Colorado. The organizers of the session expected two hundred attendees; two thousand showed. Impressed by student interest and by each other, Jack and Joseph began to collaborate. By 1976, joined by Sharon and supportive friends, they purchased a former Catholic seminary in Barre, Massachusetts, and founded the Insight Meditation Society. Over the last thirty years, IMS teachers have introduced vipassana to tens of thousands of students in courses that usually last between one and ten days. IMS also offers an annual three-month course, the "rains retreat," patterned after the Buddha's practice of using India's monsoon season to withdraw his monks from the world and restore them through three months of full-time meditation.

Jack moved to California in 1984 and with others founded the Spirit Rock Meditation Center in Woodacre,

twenty-five miles from San Francisco. Though closely affiliated with IMS, it has been more experimental and eclectic.
Jack's training in clinical psychology led him to believe that
not all of the people drawn to Buddhist meditation are right
for it and that some of them require supplemental strategies.
Therefore, although vipassana remains the core of the Spirit
Rock curriculum, it also hosts spiritual teachers from other
traditions and a variety of conferences and workshops. To
make the Dharma more available and relevant to more
people in the midst of their busy work and family lives, Spirit
Rock has experimented with meditation retreats of shorter
duration, "sandwich retreats" of two weekends and the week-
nights in between, and a family program, which offers instruction to children, teens, and families. Whereas many
centers have limited themselves mainly to offering solitary retreats, Spirit Rock has paid considerable attention to creating
community. It has also played a key role in advancing the
American inter-Buddhist dialogue by hosting many ecumenical gatherings including, in 2000, a historic, large meeting of
Western Buddhist teachers hailing from many traditions.[5]

On the other side of the American vipassana fence stands
the remarkable S. N. Goenka, whose "role in disseminating
the practice of *vipassana* world-wide," says Buddhist historian Stephen Batchelor, "is unparalleled."[6] Goenka-ji, as he is
called, was born into a Hindu business family in Myanmar
(Burma) some seventy-eight years ago. A wealthy industrial-
ist and civic leader by his mid-twenties, he began to suffer
from headaches that neither Asian nor Western doctors could
cure. To his devout Hindu family's dismay, he signed up for a
Buddhist meditation course in the desperate hope it might
bring some relief. By the course's end Goenka-ji had discovered something far greater than a cure for his headaches. His
life had changed.

The course that effected this change was taught by U Ba Khin, a Burmese layman and government official who had founded the International Meditation Center in Yangon (Rangoon) and developed the ten-day retreat schedule that Goenka-ji would make widely known. Goenka-ji continued to study with U Ba Khin for fourteen years, during which time his old life fell from him "like a worn-out skin."[7] He dedicated his new life to teaching the Buddha's ancient Eightfold Path and disseminating vipassana meditation. In 1969 he immigrated to India and began to teach there.

His first meditation course attracted only a handful of people, including his parents. Other courses, offered here and there in rented facilities, soon attracted thousands of people a year from every section of Indian society. Hundreds of young European, Australian, and American travelers in India found their way into a "Goenka course." Part of the attraction was general: a chance to practice an authentic spiritual discipline on the fabled ground of Mother India. Another part, however, was specific: it was Goenka-ji himself—not as a guru to revere, but as a teacher extraordinaire. His mastery of ancient Pali, the Buddha's tongue, brought his teachings to life. His remarkably nuanced English brought a crisp clarity to daily instructions. And his evening talks, blending humor, encouragement, and exposition, revealed the elegant symmetry between vipassana practice and Buddhist doctrines such as the Four Noble Truths, the Eightfold Path, and the three marks of existence.

By 1976 Goenka-ji had founded the Vipassana International Academy in Igatpuri, north of Bombay, the first of many Indian centers. In 1979, at the request of many of his former students now back home in the West, he began regular teaching visits abroad. Since then his American students have founded centers in Shelburne Falls, Massachusetts;

North Fork, California; Kaufman, Texas; and Ethel, Washington. Courses in this lineage currently serve an estimated one hundred thousand students annually at over eighty centers around the world. Since Goenka-ji began to teach, some half million people have sat vipassana courses under his tutelage, an estimated thirty-five thousand in America. Special vipassana programs are also provided to drug addicts, prison inmates, street children, and other groups, both in India and abroad.

Every one of these courses is offered free of charge. Goenka-ji's position is that since the Buddha taught vipassana for free, so should we. Donations are accepted and encouraged, but they can come only from people who have actually taken at least one course, experienced benefit, and wish to make such benefit possible for others.

Goenka-ji's teaching fully shares the posttraditional appearance of the rest of American vipassana. Gone are the robes, the rituals, and other accoutrements of Buddhism in its various Asian cultural manifestations. For all intents and purposes, only the ethics and practical psychology of the Eightfold Path remain. And though this appears "liberal," Goenka-ji would protest that his motive is highly conservative: to *return* to the Eightfold Path as it was taught by the Buddha *before* it became subject to twenty-five hundred years of cultural accretion. Still, "conservative" in Goenka's hands has strong family resemblances to the avowedly liberal wing of lay-oriented, meditation-centered, nonritualist American vipassana.

There are, to be sure, differences in the ways that the two "schools" actually practice meditation, but they will strike most observers as slight.[8] In reviewing them it must be kept in mind that when we refer to IMS–Spirit Rock practices as the "Mahasi approach," it is at best a very rough approximation meant to unify (for discussion purposes) what is in fact a broad spectrum of individual teachers' methods. Although strict uniformity of instruction is a high priority for Goenka-ji, it is not necessarily so across the way, where teachers find themselves freer to experiment. With that caveat in place, we can proceed.

First, the two schools agree that the observance of the five basic moral precepts (see Chapter 5) is the indispensable and enduring foundation for meditation. Second, in the calming-the-mind-through-concentration phase of the work, the Mahasi approach instructs the meditator to focus attention on the breath wherever it is most accessible, whereas Goenka limits the training of breath attention to the area bounded by the nostrils and the upper lip, claiming scriptural evidence that the Buddha himself taught in this way. Third, the Mahasi approach involves walking meditation—a very slow, deliberate, mindful walking for as much as an hour at stretch—as a formal part of meditation practice. Goenka-ji, however, asks that almost all of the meditation be done in the sitting posture. Mindfulness during normal walking is encouraged, but the snail's-pace kind of walking meditation is no part of this practice.

Fourth, in the insight phase of the work, the Mahasi approach emphasizes nonreactive awareness of whatever one's attention is drawn to within the four fields of mindfulness: body, body sensations, mind, and mind objects. As Fronsdal puts it: "The practitioner is taught to become aware with clear recognition, of the full range of physical, sensory, emotional, psychological and cognitive experiences."[9] Goenka-ji, on the other hand, largely limits the practice of nonreactive awareness to the field of body sensations, saying that all four fields are involved in this one. Goenka-ji also strongly emphasizes the continuous systematic movement of awareness

from head to feet and feet to head (called "sweeping"), believing that this leads to the subtlest levels of sensation
awareness and thus to a more thorough deconditioning of
the mind. Fifth and finally, the schools share the belief that
meditation leads to a progressively deep seeing (vipassana)
in one's own body and mind of the three characteristics of
all phenomena—impermanence, absence-of-self, and unsatisfactoriness—a seeing that undoes the conditioning in which
ignorance, craving, and aversion have ensnared us and leads
by degrees to the inner freedom called nibbana (nirvana).

We can now address the issue that most polarizes the two
worlds of American vipassana. Bluntly put, Goenka-ji is not
much interested in experimentation or dialogue. He believes
that the method of practice he has inherited has proven its
power many times over, and he refuses to tinker with it in
any way. Across his international network of meditation centers every effort is made to keep the teaching completely uni-
form and completely free of innovation. And dialogue, the
opportunity for collaboration among Buddhist traditions that
in Asia rarely had contact with one another, does not interest
him at all. Toward other lineages his attitude seems to be
"Let a thousand flowers bloom," while he resolutely keeps
his own lineage as carefully fenced and free from cross-polli-
nation as he possibly can. In a recent interview conducted by
an Insight Meditation teacher, Goenka-ji was asked about
this stance:

Interviewer: *There is a feeling in the Western Buddhist
community that you have kept your dhamma
[dharma] groups somewhat apart from the rest of
us. For instance, last year there was a large
gathering of Western dhamma teachers from all*

*traditions and lineages but nobody representing
your group attended, even though they were invited.*

S. N. Goenka: *I am concerned that at this sort of
gathering, controversy will start, debate will start.
This will create bad feelings. People will argue, "I
am teaching like this, not like that." If this debate
doesn't happen at this meeting, it may start at the
next. Buddha said, "Debate and controversy are
harmful. When arguments arise, it is dangerous for
the dhamma."*

Interviewer: *More to the point is that many people
would love to have your sangha be a part of the
larger Buddhist community, to share with all of us
your wisdom and how you are offering the dhamma
to the Western world.*

S. N. Goenka: *I am part of Buddha's teaching. But
what can be gained from these meetings, even if we
just explain our techniques? It is better to be happy
carrying on my own way, and you carrying on your
own way. People are getting benefit. This is enough
for me.*[10]

Goenka-ji is apparently concerned that the noble intentions
of dialogue and innovation can sometimes produce the quite
unintended result of dilution or, more technically,
declension—the gradual loosening of doctrine and method
over successive generations.

Another dimension of the issue at stake here is introduced by Ajahn Amaro and Thanissaro Bhikkhu (Geoffrey
De Graff), Euro-American Theravadin monks who in some
ways stand outside the lay vipassana movement we have
been describing. Amaro was ordained in Thailand in 1979

and is now co-abbot of the Abhayaghiri Buddhist Monastery in northern California. Thanissaro was ordained in a different Thai lineage in 1976 and is now the abbot of the Metta Forest Monastery in the mountains north of San Diego, California. Motivated by the opportunity to interest Americans in the Buddhist monastic life, Amaro has taught within the Insight Meditation movement. Thanissaro, however, decided not to join the Insight Meditation group of teachers due at least in part to his conviction that Buddhist practice requires a sustained rigor not readily available within the lay retreat format. The two would appear to agree, however, that even the semimonastic rigor required of long-term lay practitioners in the Goenka lineage might not be sufficient to assure the Dharma's potency in the West. By virtue of their monastic vocations, Amaro and Thanissaro appear to agree with Robert Thurman (in Chapter 17), that an indispensable requirement for a truly strong, deeply-rooted American Buddhism will be a solid corps of professionally homeless persons—Buddhist monks and nuns.[11]

The assimilative and nonassimilative trends within American vipassana throw into sharp relief a debate that is occurring across the American Buddhist landscape. Every Buddhist knows that everything changes, including Buddhism itself, and that ways of teaching the Dharma will continue to evolve. Some will naturally wish to keep their inheritances as strict and uncompromised as possible. Others will feel just as naturally drawn toward the integration and novelty that allow the stream of Dharma to refresh humanity in new and vital ways. As Buddhist historian Steven Batchelor has aptly said: "If Buddhism is to survive in the West, it has to avoid the twin dangers of excessive rigidity, which will lead to marginalization and irrelevance, and excessive flexibility, which will lead to absorption by other disciplines and a loss of distinctive identity."[12] Here, under the magnifying glass of American Buddhism, we may be witnessing what may prove to be the besetting issue that multiculturalism is posing for religion at large in the twenty-first century.

❧❧❧

THE FLOWERING OF FAITH:
*Buddhism's Pure Land Tradition**

On first thought it seems curious to have come to the end of this book as we had envisioned it and realize that we have given only a passing nod to the Pure Land tradition, the presence of which is found throughout the whole of Mahayana Buddhism. But now that this oversight has surfaced we see that it comes with a ready explanation all but built into it.

With the Maoist/Marxist, militantly atheistic takeover of China, China became for thirty years, 1950–80, a hole in the world's religious map, and into the third millennium that hole is still only thinly papered over, for though religions are now officially tolerated, China's party line remains atheistic. Thus, in the half-century in which American interest in Buddhism burgeoned, the rivers that fed that interest skirted China. The river that flanked China on the south was Theravadin and uninterested in the Pure Land, while the river that bounds China on the east flowed from the land that the United States all but

---

*This Afterword would not have been possible without the inestimable help of the Shin Buddhist minister Rev. Tetsuo Unno. The authors are deeply grateful to him for his unflagging support and learned council, all graciously proffered.

fell in love with after WWII—Japan. And while it is true that, for example, the largest sect in Japanese Buddhism is Shin Buddhism, which is in the Pure Land tradition, America wasn't interested in either Pure Land or Shin because they both appeared to be mirror images of Christian monotheism. It was Zen that stood out as different—intriguingly and captivatingly so in the descriptions of Daisetz[1] Suzuki who almost single-handedly brought Zen to the West. Eugen Herrigel's work from the early 1950, *Zen in the Art of Archery,* set off a seemingly unending series of similar works with "Zen and" in their titles. Notable among these was Robert Pirsig's philosophically oriented *Zen and the Art of Motorcycle Maintenance* which commanded a huge audience and was followed by lighter fare—ranging from *Zen and the Art of Management* to outright spoofs such as *Zen and the Art of Tennis*— whose number rose to half a hundred.

Zen also figured in the emergence of the Beat Generation which included writers and poets such as Jack Kerouac, Allen Ginsberg, and Gary Snyder, all of whom were attracted to (and to some degree involved with) Zen. In the atmosphere of those days there was no chance for Pure Land Buddhism to get a word in edgewise. Pure Land entered America quietly through Japanese immigrants who founded their churches and kept their religion pretty much to themselves.

Having now acknowledged our having neglecting Pure Land Buddhism, and having added to this acknowledgment an excuse for this neglect, the authors are resolved at this last moment to try in the space available to make amends. Because Smith was born and raised in Pure Land country—a Chinese town inland from Shanghai—and was privileged to enjoy a twenty-year friendship with Daisetz Suzuki, whose name is going to reappear immediately in a surprising context, Novak is turning this Afterword over to him.

• • •

In the West Daisetz Suzuki will probably always be remembered as the man who brought Zen to America almost by himself, for this is the side of him that he disclosed in his English writings. All the while, however, in books for his own people (which were not translated into English until after his death) he was sounding, deeper and ever more deeply, the writings of the Pure Land school in order to argue the essential unity of Mahayana Buddhism as a whole. But to a West that has been exposed almost entirely to Suzuki's Zen writings it will come as a surprise to find him writing that "of all the developments Mahayana Buddhism has achieved in the Far East, the most remarkable one is, according to my judgment, the Shin teaching of the Pure Land school."[2]

What is that teaching, and more basically, what is that school? As was said at its opening, this Afterword is required because Pure Land Buddhism was overlooked, and (as was also noted) one of the reasons for the oversight is that when Westerners began to be interested in Buddhism, its Pure Land school looked too much like Christianity to seem interesting. But now a curious twist kicks in. Once we *do* take an interest in it, that which put the West off turns out to be the best way to get into it. Not to get to its heart—a point that cannot be overemphasized—but to get our understanding of it underway.

The strand of Christianity that serves best here is the one that bears the signatures of Saint Paul and Martin Luther. These are not household names in contemporary, secularized, and multicultural America, but the essence of their teachings has worked its way into American culture at large. One thinks of the way "Amazing Grace" with its key line, "that saved a wretch like me," has almost become an American folk song. And if we glance back at the years of the American frontier, the linchpin of the evangelists' message to the pioneers in those times was likewise God's redeeming grace. Biographies of perhaps the most famous of these evangelists, George Whitefield, show him riding into frontier towns announcing "I smell hell," and proceeding for the duration of his stay to wrestle with village drunkards as if an angel of the Lord were wrestling with gutters-full of Jacobs. And mention of drunks takes us to Alcoholics Anonymous and its Twelve Step program, which requires that the alcoholic face unflinchingly the fact that he is defeated. He must give up all thought that someday, somehow, he will be able to conquer his affliction on his own and, having rid himself of that illusion, turn his life over to a "higher power."

I am tempted to say that all that is necessary for understanding Pure Land Buddhism is to see it as a case of this syndrome working its way out under Asian skies and dialects, but of course that would be inaccurate. Pure Land Buddhism is much more than grace in Asian vernacular, but as I say, the above paragraph does point us in the right direction. As Chapter 7 of this book recounted, the disagreement over whether nirvana is to be won through self-effort or grace was an important factor in causing Buddhism to split into Theravada and Mahayana, and what was to become Pure Land took the lead in arguing that Other-power does more to get us to enlightenment than does self-power. The dichotomy was never categorical; Theravadins granted that implicitly if not explicitly the Buddha's teachings allowed some room for grace. To cite one concrete instance, there is much bowing in both branches of Buddhism, and bowing one's head is an age-old gesture of laying down the "I" as a sign of respect and reverence for something that is perceived to be greater than the "I." The Mahayanists, though, argued that to some disciples the Buddha disclosed that the universe is at heart a veritable factory for Buddha-making. Because the universe is unthinkably old and because liberation has always been possible in it, liberation has been attained innumerable times during its countless eons, most recently by the Buddha of and for the present age, Siddhartha Gautama. The efforts and enlightenments of all these Buddhas, not to mention the incredible efforts of countless bodhisattvas, have together produced nothing short of an infinite treasury of merit, a storehouse of salvific energy personified by Buddhas and bodhisattvas who dwell in innumerable ethereal realms. From these luminous, heavenly beings, faithful Buddhists can request and receive, not just *some* help or help that matches their own efforts, but unlimited help. It is a Cosmic Resource that is not of human making, but from which mortals need only ask in order to receive.

This is the backbone of Pure Land, but to make a difference in human lives, doctrines need to be decked out in story form—philosophy comes later—and Pure Land's presiding story runs like this:

Eons ago a monk named Dharmakara made forty-eight unsurpassed vows, which came to be known collectively as the Primal Vow. Of these, the Eighteenth Vow was the decisive one. "When I have attained Buddhahood," Dharmakara vowed, "if all beings, trusting in me with the most sincere heart, should wish to the born in my country (the Western Paradise that he vowed he would create) and should utter my Name one to ten times, and if they should not be born there, may I not attain enlightenment." But the Buddha for our age, Shakyamuni, had assured his disciples in his famous sermon on Vulture Peak that Dharmakara in fact had attained enlightenment, so everyone who sincerely utters Dharmakara's name will be born in his Western Paradise.

Descriptions of the Western Paradise are nothing if not extravagant. According to Mahayanist legend, Gautama told his favorite disciple, Ananda,

Now, O Ananda, that world called Pure Land is
prosperous, rich, good to live in, fertile, lovely and
filled with many gods and men. O Ananda, that Pure
Land is fragrant with sweet-smelling scents, rich in
manifold flowers and fruits, adorned with gem trees,
and frequented by tribes of manifold sweet-voiced
birds.

   There are lotus flowers there as much as ten yojanas
[ninety miles] in circumference. And from each lotus
there proceed thirty-seven hundred thousand million
rays of light. And from each ray of light there proceed
thirty-seven hundred Buddhas with bodies of golden
color, who go and teach the Dharma to beings in the
immeasurable and innumerable worlds of the Pure
Land.[3]

And on and, on, with music and heavenly pleasures of
every conceivable sort colorfully described. We get the idea.
It all seemingly appears to be very much like a Buddhist version of the popular Christian notion of heaven bedecked
with pearly gates and streets of gold.

Regarding Dharmakara, suffice it to note that this monk,
who became a bodhisattva and went on to achieve Buddhahood, came to be known as Amitayus and Amitabha. The
former means Infinite Life, signifying Boundless Compassion
and the latter, Infinite Light, signifying Boundless Wisdom.
In Japanese context, the two became melded into the one
term, "Amida." The practice of orally saying Amida's name
is known as "Nembutsu," which is made up of the six characters "Na-Mu-A-Mi-Da-Butsu," or "Namu-Amida-Butsu."[4]
"Namu" means "to take refuge in" or "to entrust oneself
wholly to," and Amida-Butsu is the Buddha who is the object of that trust.

As the idea of the Nembutsu evolved it became, as with Shinran, an expression of gratitude for being granted entrance into that spiritual stage in which the entruster is absolutely assured of not regressing to previous states of faithlessness, ignorance, and suffering, and of being born into the Pure Land from which entrance into nirvana will be easy.

This with minor variations is the storyline of Pure Land Buddhism, and as is often the case with stories, philosophical depths lie hidden beneath the surface. The Pure Land story would not have flourished as widely and for as long as it has, had it not contained features that philosophers found pointing to the deep nature of things. In the case of Pure Land, the most influential of the deep thinkers was Nagarjuna, who is revered as the first of its Seven Patriarchs. His Middle Way *(Madhyamika)* philosophy of *sunyata* (emptiness) serves as probably the pivotal philosophical foundation of Pure Land teachings, especially as explicated by the Third Patriarch, T'an-lun.

A third century C.E. brahmin who had converted to Buddhism, Nagarjuna's towering achievement was to carry to its logical extreme the Buddha's doctrine of anatta, that all things, because conditioned by all other things, lack self-existence and are empty of own-being. What needs to be spelled out here is how Nagarjuna's insight into emptiness paves the way for Pure Land's vision of a grace-filled universe.

The Buddha taught that beings, confused as they are by ignorant desires and fears, are caught in a vicious cycle called samsara, freedom from which—nirvana—was the highest human end. As a loyal disciple of the Buddha, Nagarjuna affirmed this; but he also found many Buddhists striving for a nirvana that seemed infinitely distant, the reason being that they were subtly and corrosively attached to two ideas: the first, that nirvana is an ultimate "thing" to be attained; and second, that it is to be attained by individual egos. Nagarjuna saw both ideas as failures to understand the Buddha's key teaching that *everything* is empty of own-being.

Nagarjuna's astonishing inference was that since *both* samsara and nirvana are empty of own-being, they are in the last analysis one. In his own words, "there is not the slightest difference between the two." They are not two separate realities, but one vast field of empty arisings seen either through a veil of ignorance (and thus binding) or in the light of wisdom (and thus freeing). We may still *speak* of a "path" from samsara to nirvana, but only as a provisional truth. The ultimate truth is that nirvana is not infinitely distant but infinitely near, reaching gracefully toward us, as it were, and being the ground on which we already stand if we but knew this. Only the blinders of egoism hide this truth from us.

From this vantage point, other Mahayana thinkers could now see their way clear to the Pure Land. For if the great blindness is egoism, then *any means* that heals that blindness is a skillful means (upaya) for attaining Enlightenment. And since most people find personified truths more efficacious, a Pure Land devotee's worship of Amitabha Buddha through the Nembutsu is the best way for them to internalize Nagarjuna's insight.

Having noted Pure Land's storyline and the philosophical ideas that undergird it, we can turn now to what has always concerned Buddhist teachers most, its practical applications for lived life. How can we get to the Pure Land, which is tantamount to enlightenment? Cutting through the interminable issues of which master said this and which said that, it will best serve the purposes of this Afterword if we let the net yield of Pure Land's thinking come to us through a single voice, and in the context of this book, the logical candidate for that voice is that of the great thirteenth century spiritual genius, Shinran (1173–1263). For Pure Land Buddhism has entered America almost exclusively from Japan, and the church Shinran founded is the largest Pure Land presence on this continent. The name of his church is Jodo Shinshu, where *Shin* means "true," *Shu* means "sect," and *Jodo* is the Japanese word for Pure Land. Thus Jodo Shinshu means "The True Pure Land Sect."

At the age of nine, Shinran entered the leading monastery of his day resolving to settle for nothing less than to become enlightened in his present lifetime. For twenty years he poured his entire life into realizing this objective, only to find that what he had hoped to build into an unstoppable force— his will—ran into an immovable object, his ego, filled as it was with bottomless ignorance, greed, and self-centeredness. The discovery produced a severe crisis. How was Buddhahood possible if human life has this contradiction built into it? The answer came to him from his mentor, Honen. If the gap between the finite and the Infinite is to be overcome, it is the Infinite that must take the initiative, for the incommensurable difference between the two precludes the possibility that the finite might bridge the gap with its own resources, an attempt that would be like trying to lift oneself with one's own bootstraps. If there is to be a seeking of enlightenment by the unenlightened, it must be enlightenment that serves as the real agent of that seeking, while being at the same time its object. Enter Dharmakara's Primal Vow, which assures us that the wooing has been going on for eons and is solidly in place. Once again the Christian parallel is so obvious that it almost forces itself on us, for every Sunday Christians will be found singing a hymn confessing that "I was sinking deep in sin, far from the peaceful shore,/Very deeply stained within, sinking to rise no more," and on to the chorus, which announces the way out: "Love lifted me, love lifted me,/When nothing else could help, love lifted me." In the case of Pure Land, all the seeker need do is accept Dharmakara's compassionate overture by pronouncing trustingly his name, Amida in Japanese, or in full, "Namu-Amida-Butsu."

This simple storyline with its almost pat formula for reaching Enlightenment could not have produced geniuses as great as Honen and Shinran had these giants not also been world-class metaphysicians who heard in the formula reverberations of eternity. They culled from the writings of the great Buddhist masters with a thoroughness few others have equaled, and the gist of what they came up with (presented, as I say, primarily in Shinran's voice) runs roughly like this:

Ultimately all religions are paradoxical. (Parenthetically, they have to be, for if quantum mechanics and relativity theory cannot be consistently explained in ordinary language any more than our three-dimensional planet can be accurately drawn on the two-dimensional pages of geography books, it is inconceivable that Reality, which includes the quantum and sidereal worlds, is less mysterious than they are.) The towering paradox that religion confronts us with is its insistence that the opposites that texture the world we normally experience are, when rightly understood, actually one. This is the famous doctrine of the *coincidentia oppositorum,* the coincidence of opposites. Addressing this point, Shinran argued that we have not gotten to the root of things until we perceive the Absolute Unity that is their foundation. The consequences of this perception (as Nagarjuna saw) are momentous, for if we take them seriously they pull us up short with realizations like these: Our "abysmal sinfulness" is identical with the Unthinkable Power that saves us. The "I" in the I-Thou relationship is identical with the "Thou" and the finite "I am" is identical with the "Eternal I Am." We utter the Nembutsu but really Amida utters it through us.

Our longing to become eternal is actually Amida's longing that we become so. Our love for Amida is his love for us. The indwelling Light in our hearts is His Holy Light. Our dying to our finite selves is our rebirth in the Universal Self. Our passivity is Tireless Activity working within us. This sordid world *is* the Pure Land. And on and on, with every dichotomy collapsing into—what shall we call it? Amida's Infinite Light and Compassion? The Primal Vow? The Nembutsu? Eventually words give out and we are left with the wordless apprehension that we are always already free. Life's trials have been surmounted and we are at peace.

In its profoundest reading, that last sentence stands, but it is impossible to keep it in focus because the everyday world claws at us and we are forced to deal with its omnipresent subject-object dichotomies. But once one understands that their limiting oppositions are only provisional, one sees them as staking out a playing field for compassionate action. Wisdom—which can never be separated from its correlate, Compassion—consists in seeing everything we encounter as a candidate for us to respond to compassionately. And more. We see the things that come toward us as being as compassionate as we are for offering themselves to us as stepping stones for our spiritual development if we respond to them compassionately. They are outlets through which our essential nature can find expression and work its way into our everyday understanding of ourselves.

In the long years that followed his switch from self-power to Other-power (Shinran died at the age of 90), he devoted himself to deepening his understanding of the Primal Vow, working the implications of that understanding into his daily life, and tirelessly transmitting what he had come upon to the throngs of disciples that were rapidly drawn to him. There is no doubt that the speed with which his message caught on was in large part due to the fact that he and Honen directed their teachings primarily to the lower classes. Zen had always been a rather aristocratic tradition, attracting first the nobility and later the leaders of the warrior (samurai) class. It emphasized strenuous effort, and only the well-to-do had the time for it. But Honen and Shinran lived in a time of great turbulence. A string of natural calamities was accompanied by power grabs by warring factions that were tearing the islands to shreds. In such times the lower classes had to struggle desperately just to stay alive. Shinran's mentor and Shinran himself sought out those masses, preaching to people who started working when the stars were still shining in the skies and continued long after dark to return to collapse like exhausted beasts of burden on beds of straw. The only thing that could induce such people to steal a few moments from their sleep depravation was hope, and Honen and Shinran worked themselves to the bone to bring it to them. To people who, with neither time nor energy for spiritual exercises had resigned themselves to the hell of unending incarnations of indentured servitude, these new teachers brought the life-reviving message that self-help is not the only way! It is not difficult to imagine how sweet the news of the Primal Vow sounded to them. The Nembutsu is free and easy, they were being told, much like taking a pleasant boat ride.

To aid their converts, Honen and Shinran filled temples with paintings and statues of Amida Buddha. Pure Land Buddhism has always favored the former of these, while positioning sound above visualizations of any sort. The Nembutsu can be (and often is) written, but Pure Land Buddhism holds out for hearing as the most effective way for the Primal Vow to enter the human heart. The hearing can be of a priest's sermon or at a deeper, mystical level, hearing the power of Amida as Light.

Realization of the existential consequences of Dharmakara's

Vow can strike as suddenly as a lightning bolt, but lightning is not the same as abiding daylight and to convert it into such requires spiritual practices. The attainment of Wisdom/Enlightenment is the aim of Buddhist life, but it cannot be said too often that the crucial issue is how to reach it. What should we *do* to purge our lives of blind passion and inverted views. Pure Land Buddhism's answers is: deepen our entrustment to the Primal Vow and hold fast to that trust in the face of life's trials and afflictions. There are times in life where the line between being and nonbeing grows thin. The moment of death. A birth or car accident. Sudden joy, or even sudden grief. All of these can trigger the realization that Dharmakara's Primal Vow is always with us, dogging every step of our life's journey, imploring us to look beyond our trivial concerns and *see*. And be. We can be grateful for these spontaneous incursions, but for the long haul nothing is as efficacious as the Nembutsu. Honen repeated it thousands of times each day, while Shinran placed ultimate emphasis on the depth of the practitioner's entrustment; if the Nembutsu is said with absolute entrustment, a single utterance will suffice. Shinran also implied that the Nembutsu should be uttered as if the Amida made his Primal Vow solely for the individual who at the moment is uttering it.

At every turn Shinran placed ultimate emphasis on the here and now. It is only right now, in the present streaming moment, that we are alive or dead, stagnant or growing, sinful or filled with the wisdom and compassion of the Primal Vow. In every instant we have the choice to turn our hearts to the Vow or give in to greed, sloth, and impulses, idly clicking through TV channels and filling our hours with the distractions of practical duties. And, of course, forgetfulness dogs us constantly. We lose sight of the fact that the Gift beyond all gifts is absolutely One with the Primal Vow. And when our self-power—our power to reason and calculate our way to resolutions—fails, the Primal Vow surfaces and manifests itself with and in each utterance of the name of the Amida Buddha.

As this Afterword reaches its close, I want to stress again that most of the terms and concept that I have placed in Honen and Shinran's mouths can be found sprinkled throughout the vast corpus of Mahayana texts, most of which have not yet been translated into English. Thus the present account provides a mere hint of the wisdom that the Pure Land tradition embodies.

That said, if I were to attempt to summarize the message Honen, and in particular Shinran through Jodo Shinshu, offered the world, it might run as follows:

Should those who seek Enlightenment—with its erasure of desire and the ego—find it impossible to do so, salvation and eventual Enlightenment are still possible. Not through one's own efforts, but on the contrary through abandoning those efforts and entrusting themselves to a Power-other-than-one's-own, the Other-Power.

In terms of a Being, this means entrusting oneself to Amida Buddha, who is Infinite Wisdom and Compassion, and, concomitantly being made one with him. As an act, it means hearing, truly hearing, Amida's Primal Vow, which assures the unconditional salvation of those who appear to be hopelessly beyond its pale. And orally, it means to utter Amida's name, Namu-Amida-Butsu, out of a sense of boundless gratitude. Not gratitude as an act of virtue, but as the outpouring, as spontaneous as a birdsong, of a being who in human terms is utterly unsaveable, and yet in reality is saved by the virtue of the Other Power; that is, by the Boundless Compassion and Wisdom of the Amida Buddha.

NOTES

CHAPTER 1

1. The word in the case of Jesus was different, but the direction of the question was the same.

2. Buddhist societies distinguish between two world-moving wheels, the wheel of political power *(anachakra)* and the wheel of moral-spiritual power *(dharmachakra)*. The Buddha's followers honored him with the title *Chakravartin* because he turned the *latter* wheel, setting it rolling over the earth for the awakening and benefit of all beings. The similar double meaning of *messiah* plays a role in the story of Jesus.

3. Adapted from Edward Conze's translation of *The Buddhacarita (Acts of the Buddha),* in E. Conze, *Buddhist Scriptures* (London: Penguin, 1959), 44–45.

4. See Clarence H. Hamilton, *Buddhism: A Religion of Infinite Compassion* (1952; reprint, New York: Liberal Arts Press, 1954), 14–15.

5. A watch is a four-hour period. The first watch of the night stretches from dusk to 10 P.M.

6. "Who sees Dependent Arising sees the Teaching; who sees the Teaching sees Dependent Arising." *Majjhima Nikaya,* Sutta 28. For an explanation of dependent arising, see Chapter 3, note 14.

7. *Dhammapada,* 153–54, authors' free rendering.

8. See Hamilton, *Buddhism,* 3–4.

CHAPTER 2

1. Literally, "Birth Stories," i.e., moral tales of the former lives of the Buddha. A major genre of Buddhist literature, the *Jataka Tales* recount 546 previous lives of the Buddha, 357 as a human being, 66 as a god, and 123 as an animal.

2. Quoted from the *Digha Nikaya*, in J. B. Pratt, *The Pilgrimage of Buddhism and a Buddhist Pilgrimage* (New York: AMS Press, 1928), 10.

3. Related in Pratt, *Pilgrimage,* 12.

4. Quoted in Pratt, *Pilgrimage,* 8.

5. Quoted in Pratt, *Pilgrimage,* 9.

6. Quoted in Pratt, *Pilgrimage,* 10.

7. That is, one who has traveled the Way to its end.

8. *Majjhima Nikaya,* Sutta 72, quoted in Pratt, *Pilgrimage,* 13.

CHAPTER 3

1. William James, *The Varieties of Religious Experience* (New York: Macmillan, 1961).

2. Quoted in B. L. Suzuki, *Mahayana Buddhism* (1948; rev. ed., London: Allen and Unwin, 1981), 2.

3. Soma Thera, trans., *Kalama Sutta,* in *The Wheel,* no. 8 (Kandy, Sri Lanka: Buddhist Publication Society, 1965).

4. E. A. Burtt, *The Teachings of the Compassionate Buddha* (New York: Mentor Books, 1955), 49–50.

5. *Samyutta Nikaya* 42, 6, quoted in Georg Grimm, *The Doctrine of the Buddha* (Delhi: Motilal Banarsidas, 1958), 54.

6. The Ten Fetters are soul belief, chronic doubt, clinging to mere rules and rituals, sensuous craving, ill will, craving for fine-material existence, craving for immaterial existence, conceit, restlessness, and ignorance.

7. Burtt, *Teachings,* 18.

8. See, for example, Burtt, *Teachings,* 32.

9. We have paraphrased slightly the discourse as it appears in *Majjhima Nikaya,* Sutta 63, as translated by E. J. Thomas in *Early Buddhist Scriptures* (New York: AMS Press, 1935), 64–67.

10. Quoted in F. L. Woodward, *Some Sayings of the Buddha* (London: Gordon Press, 1939), 283.

11. Quoted in Burtt, *Teachings,* 50.

12. Quoted in Christmas Humphreys, *Buddhism* (Harmondsworth, England: Pelican Books, 1951), 120.

13. Quoted in Woodward, *Some Sayings,* 223.

14. *Samyutta Nikaya* 2, 64–65, as quoted in A. Coomaraswamy, *Hinduism and Buddhism* (New York: Philosophical Library, 1943), 62.

The longer version of dependent arising is expressed as a set of twelve interlocking conditions *(nidana):* (1) Ignorance (of *anatta* [no-self] and the Four Noble Truths) occasions (2) dispositional tendencies, which occasion (3) consciousness, which occasions (4) name and form, which occasion (5) the six sense fields, which occasion (6) contact between our senses and external reality, which occasions (7) sensations in the body and mind, which occasion (8) the entire habit structure of wanting and not wanting, which occasion (9) clinging, which occasions (10) becoming, which occasions (11) rebirth, upon which necessarily follow (12) illness, decay, death, and all their related suffering. Eradicate ignorance, says the Buddha, and the conditions of our bondage begin to fall like dominoes. Using these twelve conditions to explain dependent arising, however, has been likened to using the collision of a few bowling balls to explain the interactions of subatomic particles. The matter is far subtler. The all-encompassing range of dependent arising is best caught in the shorter, though deceptively simple formulation just alluded to: "When this is, that is; this arising, that arises. When this is not, that is not; this ceasing, that ceases." This latter formulation helps us to see that dependent arising pertains not only to the human personality, but to the whole of reality. *All* things and events depend for their existence on other things and/or events—which resembles field theory in physics and entails as its corollary that all things and events are empty-of-own-being *(anatta).*

15. Woodward, *Some Sayings,* 294.

16. The Buddha's social vision is expertly assessed in Trevor Ling's neglected gem *The Buddha: Buddhist Civilization in India and Ceylon* (New York: Scribner, 1973).

17. *Samyutta Nikaya* 3, 18.

18. Burtt, *Teachings,* 49. As we have cited this famous charge a number of times, we should also guard against a possible misunderstanding of it. By telling his fellows to work our their own salvation, the Buddha did not mean "make your own Truth," but, rather, "make the Truth your own."

CHAPTER 4

1. Both of the central Buddhist meanings of "Dharma" (Cosmic Law and the Teaching of the Buddha) are involved in this sermon title. On the one hand, Dharma (from the Sanskrit *dhri,* to support or uphold) means Foundational Law—the Moral Order of the Universe, The Way Things Really Are. On the other hand, Dharma means the Way of living in harmony with that Law as taught by the Buddha. Whereas Dharma as Eternal Law precedes the Buddha, it is Dharma as Way that

the Buddha's teaching sets in motion, the "wheel" being an ancient, obvious metaphor for movement. (In the Buddha's time powerful people were called wheel-turners; see Chapter 1, note 2). The core of the Buddha's teaching, the Eightfold Path, came to be symbolized as an eight-spoked wheel, which in turn became an icon of the Buddha's entire teaching. The Wheel of the Dharma abides at the center of India's national flag.

2. Sir Edwin Arnold, *The Light of Asia* (1879; reprint, Los Angeles: Theosophy Co., 1977).

3. Robert Penn Warren, *Brother to Dragons* (New York: Random House, 1979).

4. Sigmund Freud, *General Introduction to Psychoanalysis* (New York: Liverwright, 1935), 344.

5. For a superb contemporary philosophical reflection on the "self that is not there," see David Loy, *Lack and Transcendence: The Problem of Death and Life in Psychotherapy, Existentialism and Buddhism* (Atlantic Highlands, NJ: Humanities Press, 1996).

6. In his first sermon, "Setting-in-Motion the Wheel of Dharma."

7. Christmas Humphreys, *Buddhism* (Harmondsworth, England: Pelican Books, 1951), 91.

CHAPTER 5

1. E. Easwaren, trans., *Dhammapada,* vv. 61, 76, 207–8 (Tomales, CA: Nilgiri Press, 1985).

2. Relating his field experience with the forest monks of Sri Lanka, Buddhist scholar Michael Carrithers says that, though he approached his subjects under the assumption that, for them, morality was merely preparatory to meditation, he came to see that it was quite the other way around. He found that they gave an "axiomatically fundamental place to *sila* (morality)," that moral cultivation was their lives' "prime purpose," and that "the monks place moral purity in the central position I had wished to accord meditative experience." See *The Forest Monks of Sri Lanka* (Delhi: Oxford University Press, 1983), 18–20. See also Philip Novak, "Mysticism, Enlightenment and Morality," in *ReVision* 12, no. 1 (summer 1989): 45–49.

3. Quoted in J. B. Pratt, *The Pilgrimage of Buddhism and a Buddhist Pilgrimage* (New York: AMS Press, 1928), 40.

4. Nyanaponika Thera, trans., *Anguttara Nikaya: An Anthology, Part II,* in *The Wheel,* nos. 208–11 (Kandy, Sri Lanka: Buddhist Publication Society), 56ff.

5. Edward Conze, *Buddhist Meditation* (New York: Harper & Row, 1975), 11.

6. Adapted from various translations.

7. The three insights described in this paragraph are insights into the *three marks of existence*—impermanence *(anicca)*, lack of self-existence *(anatta)*, and unsatisfactoriness *(dukkha)*—which are further discussed in Chapters 6, 8, and 18.

8. Ignorance here is not lack of formal education, but rather lack of insight into *anatta* (no-self, no-soul), to be discussed in Chapter 6. For Buddhists, delusion about what we really are is the "original sin," the error that leads to personalities built of craving and aversion—the warp and woof of suffering.

CHAPTER 6

1. *Samyutta Nikaya*, 38, 1.

2. Precisely this indescribable character of nirvana caused later Buddhists to speak of it as *shunyata*, or "emptiness." It is void, but not in the absolute sense. Rather, it is *de*void of finite, specifiable features, in something of the way the suprasonic is lacking in sounds our ears can register.

3. *Milindapanha*, 271, abridged as translated in Edward Conze et al., *Buddhist Texts Through the Ages* (New York and Evanston: Harper & Row, 1964), 99–100.

4. *Iti-vuttaka*, 43; *Udana* 8, 3. See Pratt, *Pilgrimage*, 88–89, and E. A. Burtt, *The Teachings of the Compassionate Buddha* (New York: Mentor Books, 1955), 113.

5. Edward Conze, *Buddhism: Its Essence and Development* (reprint, New York: Harper & Row, 1951), 40.

6. Compare, for example, its relation to Paul Tillich's "God Above God," in *The Courage to Be* (New Haven, CT: Yale University Press, 1952), 186–90.

7. Zen master Dogen, in Heinrich Dumoulin, *A History of Zen Buddhism* (Boston: Beacon Press, 1963), 159.

8. *Vajracchedika (Diamond Sutra)*, 32.

9. This, in passing, was one of the ways in which the Buddha's understanding of reincarnation differed from that of most Hindus of his day. The standard Hindu doctrine attributed rebirth to karma, the consequences of actions set in motion during previous lives. As these actions are innumerable, innumerable lives were assumed to be needed to work off these consequences. Characteristically, the Buddha took a more psychological view. Rebirth, he maintained, was due not to karma, but to tanha. As long as the wish to be a separate self persisted, that wish would be granted. It follows that since desire is the key, it is possible to step permanently out of the cycle of rebirth whenever one wishes wholeheartedly to do so.

10. *Arhat* means "noble person." There is no difference between an arhat and a Buddha in either depth of wisdom or moral goodness. The latter title is simply reserved for one who has rediscovered the Path without the help of another.

11. From the *Majjhima Nikaya* 72, Aggi-Vacchagotta Sutta, adapted.

12. Quoted in Pratt, *Pilgrimage,* 86. We have substituted the phrase "dispositional tendencies" for the word "forces" in Pratt's text. The term being translated is *sankhara.*

13. See Chapter 3, note 14.

14. *Samyutta Nikaya* 38, 1, 3, 83–84, adapted.

15. Quoted in Pratt, *Pilgrimage,* 91.

CHAPTER 7

1. The Tibetan version holds that the Buddha explicitly preached the Mahayana doctrines, but in his "glorified body" *(sambhogakaya),* which only the most advanced disciples could perceive.

2. From Shantideva's *Bodhicaryavatara,* chap. 3, vv. 7–10, quoted in E. A. Burtt, *Teachings of the Compassionate Buddha* (New York: Mentor, 1955), 136, from L. D. Barnett, *The Path of Light* (London: John Murray), 37–94.

3. Theravadins are not without a cosmology. Traditionally, it has thirty-one levels and includes many god realms and multiple domains of bliss and punishment. But as no real aid comes from these quarters, relatively little attention is given to them. Mahayana world pictures are not only often more elaborate, but also more engaging to their beholders as sources of spiritual succor.

4. Though Mahayana honors wisdom as conducive to compassion.

5. If it seems like mixing politics and religion to say this, we should realize a point that this book, focusing as it does on metaphysics, psychology, and ethics, does not go into; namely, that the great religions entered history not so much as religions in the narrow meaning of that word, but rather as civilizations. Each staked out for its adherents an entire way of life—a life-world that encompassed not only things we now consider distinctively religious, but also regions of life that the modern world divides into economics, politics, ethics, law, art, philosophy, and education.

6. An excellent start in this direction is Trevor Ling's already mentioned *The Buddha: Buddhist Civilization in India and Ceylon* (New York: Scribner, 1973).

CHAPTER 8

1. There are actually a countless number of things that could be said, but not without violating the introductory nature of the present work. For a more thorough historical account of Theravada Buddhism, we can think of no better place to start than Richard Gombrich's *Theravada Buddhism: A Social History from Ancient Benares to Modern Colombo* (London and New York: Routledge and Kegan Paul, 1988). Gombrich focuses on Sri Lankan Theravada. For rich accounts of Theravada in Thailand and Burma, see S. J. Tambiah's *Buddhism and the Spirit Cults in Northeast Thailand* (Cambridge: Cambridge University Press, 1970), and Melford Spiro's *Buddhism and Society: A Great Tradition and Its Burmese Vicissitudes* (New York: 1970), respectively.

2. *Abhidhamma* is often translated as "higher dhamma [dharma]" in the sense of a more philosophically exact explanation of the Buddha's teachings, but it is perhaps better thought of as "about dhamma" or "after dhamma" in the sense of a post-Buddha analysis and commentary.

3. This statement was the reported outcome of a conference of monks around 100 B.C.E. who were debating whether learning or practice was the basis of the Buddha's teaching. It is quoted in Richard Gombrich, *Theravada Buddhism: A Social History from Ancient Benares to Modern Colombo* (London and New York: Routledge and Kegan Paul, 1988), 152.

4. The order of Sri Lankan *bhikkunis,* or nuns, died out in the eleventh century and to this day has not been formally reestablished (see Peter Harvey, *An Introduction to Buddhism* [Cambridge: Cambridge University Press, 1990], 222–24). Nevertheless an informal order of ordained nuns who shave their heads, take Pali names, wear monastic robes, and permanently keep eight or ten precepts (that is, three to five more than are required of the Buddhist laity) does exist, and there are today about three thousand such nuns in Sri Lanka along with about twenty thousand monks.

5. It appears that for many centuries meditation practice all but died out in the Southeast Asian lands of the Theravada tradition, and that the sangha was occupied almost exclusively with scriptural study, ritual performance, moral refinement, and the education of and assistance to the laity. What the world currently recognizes as "Theravada meditation" and "vipassana" are products of rather recent rebirths of interest in meditation that can be confidently traced back only about 150 to 250 years. See Robert H. Sharf, "Buddhist Modernism and the Rhetoric of Meditative Experience," *Numen* 42 (1995): 228–83, esp. 246–59.

6. E. Easwaren, trans., *Dhammapada* (Tomales, CA: Nilgiri Press, 1985), v. 80.

7. Rick Fields, *When the Swans Came to the Lake: A Narrative History of Buddhism in America,* 3d ed. (Boston and London: Shambhala, 1992), 370.

8. March 14, 1977. The same opinion is reflected in Nyanaponika's *The Heart of Buddhist Meditation* (New York: Samuel Weiser, 1973), 89. Nyanaponika holds that what *is* necessary for progress in vipassana is a degree of concentration known as *upacara-samadhi* ("access concentration"), which is profound but still short of the first jhana.

9. *Majjhima Nikaya* 1, 240ff., quoted in Edward J. Thomas, *The Life of Buddha* (London: Routledge and Kegan Paul, 1927), 63. In a seminal article, Buddhist scholar Robert Gimello points out that in the Theravada tradition the jhanas have "no liberative value or cognitive force in themselves" ("Mysticism and Its Contexts," in Steven Katz, ed., *Mysticism and Religious Traditions* [New York; Oxford Univeristy Press, 1983], 63).

10. Sangharakshita, *A Survey of Buddhism* (Glasgow: Windhorse Publications, [1957] 1993), 192. A contemporary dissenting voice is that of the late German nun Ayya Khema, who, while acknowledging that insight is the crucial factor, feels that jhanas have been unjustly disparaged by many vipassana teachers and seeks to reclaim jhanas as an effective vehicle of insight. See S. Batchelor, *The Awakening of the West: The Encounter of Buddhism and Western Culture* (Berkeley, CA: Parallax Press, 1994), 352.

11. *The Way of Mindfulness,* 6th revised edition (a translation of the *Satipatthana Sutta* of the *Majjhima Nikaya*), trans. Soma Thera (Kandy, Sri Lanka: Buddhist Publication Society, 1998), p.11 of online version (http://www.accesstoinsight.org).

12. *Anguttara Nikaya* 4, 45.

13. *Theragatha* 468.

14. *Anguttara Nikaya* 1.

15. *Digha Nikaya* 1, Brahmajala Sutta, quoted in William Hart, *The Art of Living: Vipassana Meditation as Taught by S. N. Goenka* (San Francisco: HarperSanFrancisco, 1987), 148, condensed.

16. Craving, aversion, lethargy and drowsiness, worry and agitation, and skeptical doubt.

17. Mindfulness, analytic application of dhamma, effort, rapture, tranquility, concentration, and equanimity.

18. *Samyutta Nikaya*, Apana Sutta, quoted in Hart, *The Art of Living,* 157.

19. *Samyutta Nikaya* 52, 9, adapted and abridged.

20. Stephen Batchelor writes: "In fact, vipassana is central to *all* forms of Buddhist meditation practice. The distinctive goal of any Buddhist contemplative tradition is a state in which inner calm (samatha) is *unified* with insight (vipassana). Over the centuries, each tradition has developed its own methods for actualizing this state. And it is in these methods that the traditions differ, not in their end objective of unified calm and insight" (*Awakening of the West,* 344). That vipassana is a goal that may be reached by many methods is true (as Batchelor says) insofar as vipassana refers only to a general goal. The problem is that vipassana also refers very specifically to a particular method, namely, the *practice of insight into the three marks of existence through awareness of (one or more of) the four fields of mindfulness* (body, body sensations, mind, mind objects). This appears to reinforce the divergence among Buddhist meditative traditions that Batchelor seeks to soften.

21. Itivuttaka Sutta, 27.

22. "The Practice of Metta," in the Sutta Nipata, 145–51, trans. Nanamoli Thera, in *The Wheel,* no. 7 (Kandy, Sri Lanka: Buddhist Publication Society, 1964), 19.

23. *Majjhima Nikaya,* Sutta no. 7, "The Simile of the Cloth," trans. Nyanaponika Thera (Kandy, Sri Lanka: Buddhist Publication Society, 1988), v. 12, adapted to eliminate repetition.

24. *Dhammapada,* 183, authors' free rendering.

25. *Udana 5,* 5, Uposatha Sutta.

CHAPTER 9

1. Author Smith assumes primary responsibility for this chapter. It was begun under the influence of Dr. D. T. Suzuki's writings and person, received its final shape from six weeks of Zen training in Kyoto during the summer of 1957—weeks that included daily *sanzen* (consultation concerning meditation) with the eminent Zen master Goto Roshi; celebration of *gematsu o-sesshin* (eight days of looking into mind and heart) with monks in the monastery of Myoshinji (Temple of the Marvelous Mind); access to the manuscripts of the Kyoto branch of the First Zen Institute of America; and a number of important conversations with its then director, Ruth Fuller Sasaki.

2. A Western professor, wishing to show that he had grasped Zen's determination to transcend forms, expressed surprise when the abbot of the temple he was visiting bowed reverently to images of the Buddha as they passed them. "I thought you were beyond such things," he said, adding, "I am. Why I would just as soon spit on these images." "Very well," said the abbot in his not quite perfect English. "You spits. I bow."

3. Because of the extent to which reason was interfering with author Smith's Zen practice, his teacher, Goto Roshi, diagnosed him as having contracted "the philosopher's disease." Immediately, though, he retracted, acknowledging that there was nothing wrong with philosophy as such; he himself had a master's degree in philosophy from one of the better Japanese universities. "However," he continued, "reason can only work with the experience that is available to it. You obviously know how to reason. What you lack is the experiential premise which makes reason wise when it is reasoned *from*. For these weeks put reason aside and work for experience."

4. The two are *Soto,* stemming from Dogen, who imported the *Ts'ao-tung* school of Ch'an from China; and *Rinzai,* the Japanese version of the *Lin-chi* school, which Eisai introduced to Japan. The former considers enlightenment a gradual process; the latter contends that it is sudden.

5. Koan use is de rigueur in Rinzai Zen. Soto uses the koan, but less centrally, and tends to emphasize the practice of zazen.

6. Author Smith was told that the shortest time on record in which a koan was solved was overnight, and the longest time was twelve years.

7. Dylan Thomas, "Light Breaks Where No Sun Shines."

Koans are actually of different types, geared to the stages in the students' progress. As the mind must work differently according to the kind of koan it is assigned, a phenomenonological description of the entire sweep of koan study would be complex. What is said here applies to early koans. Isshu Miura and Ruth Fuller Sasaki's *Zen Dust: The History of the Koan and Koan Study in Rinzai* (New York: Harcourt, Brace & World, 1966) presents a comprehensive account of koan training.

8. Miura and Sasaki, *Zen Dust,* 92. Source not given.

9. Quoted in *Cat's Yawn* (New York: First Zen Institute of America, 1947), 32.

10. Quoted in *Zen Notes* 1, no. 5 (New York: First Zen Institute of America), 1.

11. A great master, Dai Osho, reported, "I have experienced Great Satori eighteen times, and lost count of the number of small satoris I have had."

12. Miura and Sasaki, *Zen Dust.*

13. From *The Sayings of the Lay Disciple Ho.* Not published in English.

14. Abridged from D. T. Suzuki's translation, in Edward Conze, ed., *Buddhist Scriptures* (Baltimore: Penguin Books, 1973), 171–75.

15. From "Zen—A Religion," an unpublished essay by Ruth Fuller Sasaki.

CHAPTER 10

1. The word the Tibetans use to translate the Sanskrit word *vajra* is *dorje,* which literally means "chief stone" (*dorj,* "stone"; *je,* "chief").

2. Huston Smith is here describing the rituals of Gyume and Gyutö—the two highest Tantric colleges in Tibet, now in exile in India—in the latter of which he engaged in field work. For particulars relating to its exceptional chanting, see Huston Smith, "Can One Voice Sing a Chord?" *The Boston Globe,* January 26, 1969; with Kenneth Stevens, "Unique Vocal Ability of Certain Tibetan Lamas," *American Anthropologist* 69 (April 1967): 2; and with K. Stevens and R. Tomlinson, "On an Unusual Mode of Chanting by Certain Tibetan Lamas," *Journal of the Accoustical Society of America* 41 (May 1967): 5. This mode of chanting has introduced a new term into the lexicography of musicology, multiphonic chanting. It was imported from India (where the art has been lost) and was preserved in the two colleges just mentioned, which continue to survive in exile in India. Tuva throat singing resembles the Tibetan version, but Mongolian multitoned chanting is produced by a different vocal mechanism and is folk music.

CHAPTER 11

1. E. Easwaren, trans., *Dhammapada* (Tomales, CA: Nilgiri Press, 1985), v. 5.

2. Easwaren, trans., *Dhammapada,* v. 6.

3. The *Heart Sutra* and the *Diamond Sutra* are among the most celebrated portions of this literature.

4. This is a paraphrase of Nagarjuna, the great philosopher of emptiness whom the Buddhist tradition often honors as the second Buddha. His classic assertion is: "There is no difference whatsoever between nirvana and samsara; there is no difference whatsoever between samsara and nirvana" (*Mulamadhyamakakarika,* chap. 25, v. 19).

5. From Hakuin's "Song in Praise of Zazen," slightly adapted.

6. The opening line of the Bodhisattva Vow is: "Sentient beings are numberless, I vow to save them." And in classic nondualist fashion the *Diamond Sutra* of Mahayana Buddhism declares that a true bodhisattva, while working to save other beings, never forgets that there are really no other beings—no separate selves—to save.

Theravadins honor the sublime sentiments of the Bodhisattva Vow, but remain skeptical about the nondualism that serves as its philosophical backdrop. Deducing from the doctrine of emptiness that "samsara is nirvana" may be eminent logic, say the Theravadins, but if it pre-cludes the effort to actually arrive at the yonder shore, where this phrase becomes fact, it becomes instead a "mantra" in the worst sense of the word—an empty ritual. The dualities taught by the ever prag-matic Buddha (ignorance and enlightenment, good and evil, samsara and nirvana) are the very dualities of our actual, lived experience, and, Theravadins feel, their erasure, however logically compelling, risks los-ing the Path. So if Mahayanists upbraid Theravadins for resting in a provisional dualism that invites a kind of spiritual egoism ("*I* am mak-ing progress from samsara to nirvana"), Theravadins chide Ma-hayanists for a nondualism that risks presumption and complacency ("There's nothing really to do").

In the last analysis, however, "Theravada" and "Mahayana" are conceptual abstractions. Reality is individual Buddhists who, irrespec-tive of their formal affiliation, must contend with this great effort/no-effort dialectic *in their own minds*. (The Christian version would be: Work out your salvation by all means, but never forget that nothing can separate you from the all-embracing love of God.) We sometimes find ourselves wondering whether the Buddha could have intended the lasting tension between Theravada and Mahayana to engender in all his followers a mutually corrective "tacking" mechanism for true sail-ing toward the yonder shore.

CHAPTER 12

1. See N. A. Nikam and R. McKeon, *The Edicts of Asoka* (Chicago: University of Chicago Press, 1959). The works by Gombrich (1988), Ling (1973), and Robinson and Johnson (1997), cited in "Sug-gestions for Further Reading," contain excellent briefer accounts of Asoka's career.

2. *Srimad Bhagavata: The Holy Book of God*, bk. 9, chap. 21, v. 12, adapted. The author of the *Bhagavatam* is Shukadeva, but this par-ticular quotation is attributed to Rantideva.

3. Yet there is breaking news on this front, and this centuries-old Hindu doctrine now seems destined to wane. We have recently learned that on November 11, 1999, the Shankaracharya of Kanchi, a high-ranking Hindu religious authority, gave a press conference urging Hin-dus to stop teaching that the Buddha is an incarnation of Vishnu. Since then no fewer than twenty-two top Hindu religious leaders have signed a document that reflects this position. The promoter of the press con-ference and the document was S. N. Goenka (profiled in Chapter 18), an influential Buddhist meditation teacher. Goenka has called on Hindu leaders to realize that persistence in this belief is doing untold

damage to India's relations with neighboring Buddhist countries. The relevant paragraph of the historic document reads as follows: "Due to whatever reason, some literature was written (in India) in the past in which the Buddha was declared to be a reincarnation of Vishnu and various things were written about him. This was very unpleasant to the neighboring countries. In order to foster friendlier ties between the two communities we decide that whatever has happened in the past (can't be undone, but) should be forgotten and such beliefs should not be propagated" (Press Release, Maha Bodhi Society, Sarnath, India, November 11, 1999).

The reason that the doctrine "can't be undone" is because it appears in the Puranas, Hindu sacred scriptures, which simply can't be unwritten. Yet as the wider religious history of the world teaches, aspects of sacred scriptures no longer accepted as right-minded by most of the community that otherwise reveres them can be effectively neutralized simply by recalling them less and less. That appears to be the strategy embraced here.

4. According to Hindu tradition, the last of Vishnu's ten incarnations, Kalki, has not yet appeared and will do so only at the end of the Kali Yuga ("Age of Decline").

CHAPTER 13

1. The authors could not trace the source for this exact passage. A similar passage can be found in *Digha Nikaya* 16 (Mahaparinibbana Sutta), pt. 3, v. 61.

2. Quoted in Rick Fields, *When the Swans Came to the Lake: A Narrative History of Buddhism in America,* 3d ed. (Boston and London: Shambhala, 1992), 20.

3. Quoted in Fields, *Swans,* 20–21. Early Jesuit missionaries to China first embraced Buddhism, but shifted their allegiance to Confucianism when they perceived the latter as an avenue for greater influence. Their attitudes toward Buddhism were mixed.

4. Stephen Batchelor, *The Awakening of the West: The Encounter of Buddhism and Western Culture* (Berkeley, CA: Parallax Press, 1994), 167, 166.

5. See Fields, *Swans,* 47.

6. Fields, *Swans,* 25.

7. See, for example, Nietzsche's *The Antichrist,* secs. 20–22.

8. Friedrich Nietzsche, *The Antichrist,* 42. (R. J. Hollingdale, trans., *Twilight of the Idols, The Antichrist* (New York: Penguin, 1968), 154.

9. See Martin Baumann, "The Dharma Has Come West: A Survey of Recent Studies and Sources," in *Journal of Buddhist Ethics* 4 (1997).

10. Senaka Weeraratna, "The Spread of Buddhism in Germany," *Daily News*, January 31, 2001. See http://www.lanka.net/lakehouse/2001/01/31/fea05.html.

11. Among which we single out *Word of the Buddha*, a gem of essential selections from the Pali Canon.

12. For example, *The Vision of Dhamma* and *The Heart of Buddhist Meditation*. See the bibliography for complete references.

13. Batchelor, *Awakening*, 44.

14. Batchelor, *Awakening*, 114–15.

CHAPTER 14

1. Chapters 14 through 18 sketch the complex reality of Buddhism in America. We have drawn on the following excellent accounts: James W. Coleman, *The New Buddhism: The Western Transformation of An Ancient Tradition* (New York: Oxford University Press, 2001); Rick Fields, *How the Swans Came to the Lake: A Narrative History of Buddhism in America*, 3d ed. (Boston and London: Shambhala, 1992); Don Morreale, ed., *The Complete Guide to Buddhist America* (Boston and London: Shambhala, 1998); Charles Prebish, *Luminous Passage: The Practice and Study of Buddhism in America* (Berkeley: University of California Press, 1999); C. Prebish and K. Tanaka, eds., *The Faces of Buddhism in America* (Berkeley: University of California Press, 1998); Al Rapaport, comp., *Buddhism in America: Proceedings of the 1997 Conference on the Future of Buddhist Meditative Practices in the West* (Rutland, VT: Charles Tuttle, 1998); and Richard Hughes Seager, *Buddhism in America* (New York: Columbia University Press, 1999). Though focused on western Europe rather than America, also of great help have been Stephen Batchelor's superb *The Awakening of the West: The Encounter of Buddhism and Western Culture* (Berkeley, CA: Parallax Press, 1994) and Martin Baumann's "The Dharma Has Come West: A Survey of Recent Studies and Sources," *Journal of Buddhist Ethics* 4 (1997).

2. Quoted in Paul Fleischman, *Cultivating Inner Peace* (New York: Tarcher/Putnam, 1997), 133.

3. Fields, *Swans*, 62–63.

4. Fleischman, *Cultivating*, 132.

5. H. D. Thoreau, *A Week on the Concord and Merrimac Rivers*, quoted in Fields, *Swans*, 63.

6. Stephen Prothero's recent *White Buddhist: The Asian Odyssey of Henry Steel Olcott* (Bloomington: Indiana University Press, 1996) has

been hailed as an exciting portrait of this early Buddhist pioneer. Buddhist scholar Richard Gombrich believes that Olcott's *Buddhist Catechism*, an attempt to state the basic tenets to which all the world's Buddhists should be able to subscribe, represents the beginning of the modern world Buddhist movement (*Theravada Buddhism: A Social History from Ancient Benares to Modern Colombo* [London and New York: Routledge and Kegan Paul, 1988], 186).

7. In the opinion of the Buddhist historian Richard Gombrich, Dharmapala is "the most important figure in the modern history of Buddhism" (*Theravada Buddhism*, 188).

8. See Fields, *Swans*, 129.

9. Over the last two decades Asian immigrant Buddhists have been offended by the frequent suggestion that American converts, or "white Buddhists" are the chief contributors to "American Buddhism," while they have played a negligible role. Chapters 15–18, by focusing primarily on American convert Buddhism, is *not* meant to suggest that. We believe that Asian immigrant Buddhism is just as integral a part of the American tapestry as any other, and we happily report that four recent major publications on American Buddhism (by Prebish, Seager, Prebish and Tanaka [see this chapter's note 1], and D. R. Williams and C. S. Queen's *American Buddhism* [Surrey, England: Curzon Press, 1999]), have set this matter straight.

In this book, however, American Buddhism is an ancillary topic and choices had to be made. Because American convert Buddhism is the more indigenous and the more likely to interest readers of this book, we placed our focus there. We also did so because we found ourselves wanting to organize our sketch of American Buddhism around the theme of meditation, and as Charles Prebish writes in his excellent *Luminous Passage* (p. 63): "There is no disagreement among researchers that Asian immigrant communities and American convert communities engage in significantly different expression of Buddhist practice. The general consensus is that American converts gravitate toward the various meditation traditions . . . while Asian immigrants maintain practices coincident with ritual activity" (as, for example, in Pure Land Buddhism, covered in this book in the Afterword, where Asian immigrants, mostly Japanese, have clearly taken the lead). Moreover, Paul Numrich has corrected the misconception that Asian immigrant Buddhist communities attract no American converts and have no interest in meditation in his recent acclaimed study focusing on two communities in Chicago and Los Angeles, *Old Wisdom in the New World: Americanization in Two Immigrant Theravada Buddhism Temples* (Knoxville: University of Tennessee Press, 1996).

Other recent studies on ethnic Buddhist groups in North America include Penny Van Esterik, *Taking Refuge: Lao Buddhists in North America* (Tempe: Arizona State University, Program for Southeast Asian Studies, 1992), and Janet McLellan, *Many Petals of the Lotus: Five Asian Buddhist Communities in Toronto* (Toronto: University of Toronto Press, 1999).

10. The name the first Chinese Americans gave to California.

CHAPTER 15

1. James W. Coleman, *The New Buddhism: The Western Transformation of an Ancient Tradition* (New York: Oxford University Press, 2001), 119.

2. Coleman, *New Buddhism,* 14.

3. Identified as Chogyam Trungpa and Thich Nhat Hanh, in R. Robinson and W. Johnson, *The Buddhist Religion,* 4th ed. (Belmont, CA: Wadsworth, 1997), 307.

4. See, for example, Rick Fields, *How the Swans Came to the Lake: A Narrative History of Buddhism in America,* 3d ed. (Boston and London: Shambhala, 1992), chap. 16.

5. Coleman, *New Buddhism,* 15.

6. Fields, *Swans,* 369.

7. Coleman, *New Buddhism,* 223.

8. See Donald Rothberg, "Responding to the Cries of the World: Socially Engaged Buddhism in North America," in Prebish and Tanaka, *The Faces of Buddhism in America* (Berkeley: University of California Press, 1998), 269. Also, Kenneth Kraft, "Prospects of a Socially Engaged Buddhism," in Kenneth Kraft, ed., *Inner Peace, World Peace: Essays on Buddhism and Nonviolence* (Albany, NY: State University of New York Press, 1992).

9. Quoted in Rothberg, "Responding to the Cries," in Prebish and Tanaka, *Faces of Buddhism,* 268. Original source: Thich Nhat Hanh, *Peace Is Every Step* (New York: Bantam, 1991), 91.

10. Quoted in Charles Prebish, *Luminous Passage: The Practice and Study of Buddhism in America* (Berkeley: University of California Press, 1999), 109. See BPF Statement of Purpose at http://www.bpf.org/bpf.

11. Among Macy's books are *Mutual Causality in Buddhism and General Systems Theory: The Dharma of Natural Systems* (Buffalo: State University of New York Press, 1991), *Dharma and Development: Religion as Resource in the Sarvodaya Movement in Sri Lanka* (West Hartford, CT: Kumarian Press, 1983), and *Despair and Personal Power in the Nuclear Age* (Gabriola Island, BC: New Society Publishers, 1983).

12. Quoted in Stephen Batchelor, *The Awakening of the West: The Encounter of Buddhism and Western Culture* (Berkeley, CA: Parallax Press, 1994), 369.

CHAPTER 16

1. Quoted in Rick Fields, *How the Swans Came to the Lake: A Narrative History of Buddhism in America*, 3d ed. (Boston and London: Shambhala, 1992), 194. Original source: Nyogen Sensaki, *Like a Dream, Like a Fantasy: The Zen Writings of Nyogen Sensaki* (Tokyo: Japan Publications, copyright Zen Studies Society, 1978).

2. *Ashvaghosa's Discourse on the Awakening of Faith in Mahayana Buddhism* and *Outlines of Mahayana Buddhism*.

3. Quoted in Fields, *Swans*, 214.

4. Shobogenzo Zuimonki, quoted in William De Bary, ed., *The Buddhist Tradition* (New York: Vintage, 1972), 371, 373.

5. *Bendowa*, quoted in Heinrich Dumoulin, *A History of Zen Buddhism* (Boston: Beacon, 1963), 166.

6. Yuko Yukoi, *Zen Master Dogen: An Introduction with Selected Writings* (New York, Tokyo: Weatherhill, 1976), 46–47, quoted in S. Batchelor, *The Awakening of the West: The Encounter of Buddhism and Western Culture* (Berkeley, CA: Parallax Press, 1994), 129.

7. *Wind Bell 5*, no. 3 (summer, 1968): 8, quoted in Fields, *Swans*, 229.

8. Bernard Tetsugen Glassman Roshi, Dennis Genpo Merzel Roshi, Charlotte Joko Beck, Jan Chozen Bays, Gerry Shisin Wick, John Tesshin Sanderson, Alfred Jitsudo Ancheta, Charles Tenshin Fletcher, Susan Myoyu Anderson, Nicolee Jiokyo Miller, William Nyogen Yeo, and John Daido Loori.

9. Carefully profiled in Charles Prebish, *Luminous Passage: The Practice and Study of Buddhism in America* (Berkeley: University of California Press, 1999), 96–107.

10. For a fine portrait of Aitken and other American Zen masters, see Helen Tworkov, *Zen in America: Five Teachers and the Search for an American Buddhism,* rev. ed. (New York: Kodansha International, 1994).

11. James W. Coleman, *The New Buddhism: The Western Transformation of An Ancient Tradition* (New York: Oxford University Press, 2001), passim; Fields, *Swans,* chaps. 10, 11, 12, 15; Prebish, *Luminous Passage,* 8–20, 96–107; Richard Hughes Seager, *Buddhism in America* (New York: Columbia University Press, 1999), chap. 7; and Tworkov, *Zen in America.* For accounts of Zen in England, see Christmas Humphreys, *Zen Comes West: The Present and Future of Zen Buddhism*

in Britain (London: Allen and Unwin, 1960) and *Sixty Years of Buddhism in England* (London: Buddhist Society, 1968).

CHAPTER 17

1. Prof. Hopkins's work is appreciatively summarized in Donald S. Lopez's *Prisioners of Shangri-La* (Chicago: University of Chicago Press, 1998) 163–73.

2. Charles Prebish, *Luminous Passage: The Practice and Study of Buddhism in America* (Berkeley: University of California Press, 1999), 41.

3. Rick Fields, *How the Swans Came to the Lake: A Narrative History of Buddhism in America,* 3d ed. (Boston and London: Shambhala, 1992), 306.

4. Quoted in Fields, *Swans,* 307.

5. *Dzogchen,* or "great perfection," is one of the Vajrayana's many practice regimes. It refers both to a comprehensive path of training and to a specific form of meditation on that path. Traditionally regarded as a culminating effort that follows years of preparation, it has in recent years been taught sooner by a number of Tibetan teachers. Because dzogchen meditation teaches that the practitioner in some sense already *is* the great perfection she is working to achieve, it has been likened to the Zen doctrine of the oneness of practice and enlightenment (see Chapter 16) and to the *shikantaza,* or "just sitting," form of meditation associated with that doctrine. Dzogchen and shikantaza appear to be potent combinations of (1) a great effort to sustain sharp awareness and (2) a non-effort, a letting go of everything one becomes aware *of.* Here we seem to be standing on Buddhist bedrock, for a similar strategy also appears to lie at the heart of vipassana practice. Sustained awareness coupled with nonreactivity seems to be the holy discovery of Buddhist psychologists, the double-edged blade that is thought to slice through every sclerosis of conditioning that lies between bondage and freedom. A thorough account of the dzogchen path may be found in "The Innermost Essence," Chapter 10 of Sogyal Rinpoche's *The Tibetan Book of Living and Dying* (San Francisco: HarperSanFrancisco, 1992).

6. See note 5.

7. Quoted in Richard H. Seager's *Buddhism in America* (New York: Columbia University Press, 1999), 122–23. The original source is: Erik Davis, "Digital Dharma," *Wired* (August 20, 1997) [online], http://www.wired.com/wired/2.08/departments/electroshphere/dharma.html (1/8/98).

8. The three-year, three-month, and three-day retreat has been described by Ken McLeod, who has completed it twice, in Don Morreale's

The Complete Guide to Buddhist America (Boston and London: Shambhala, 1998), 229–34, and also by Rick Fields, *Swans,* 333–35.

CHAPTER 18

1. An interesting account of the putative reinvention of the Buddhist meditative tradition can be found in Robert H. Sharf's "Buddhist Modernism and the Rhetoric of Experience," in *Numen,* 42, 1995, 228–83, especially 252–59.

2. A representative list is given in Charles Prebish, *Luminous Passage: The Practice and Study of Buddhism in America* (Berkeley: University of California Press, 1999), 151–52.

3. See, for example, Richard Hughes Seager, *Buddhism in America* (New York: Columbia University Press: 1999), 249–52, for profiles of these two teachers. Denison is also profiled in Sandy Boucher, *Turning the Wheel: American Women Creating the New Buddhism* (San Francisco: HarperSanFrancisco, 1988), and Lenore Friedman, *Meetings with Remarkable Women: Buddhist Teachers in America* (London and Boston: Shambhala, 1987).

4. Gil Fronsdal, "Insight Meditation in the United States: Life, Liberty and the Pursuit of Happiness," in C. Prebish and K. Tanaka, eds., *The Faces of Buddhism in America* (Berkeley: University of California Press, 1998), 165–66.

5. The contributions of Jack Kornfield, Joseph Goldstein, and others in this domain of the American vipassana movement have been given extensive coverage in excellent recent works on American Buddhism, such as Prebish, *Luminous Passage,* 148–58; Seager, *Buddhism in America,* 146–51; and James W. Coleman, *The New Buddhism: The Western Transformation of An Ancient Tradition* (New York: Oxford University Press, 2001), 77–81, 109–13. These same works, however, say relatively little about S. N. Goenka. We are redressing this imbalance by offering a relatively brief account of the former and a more extensive account of the latter.

6. Stephen Batchelor, *The Awakening of the West: The Encounter of Buddhism and Western Culture* (Berkeley, CA: Parallax Press, 1994), 247.

7. The title of an ancient Buddhist poem.

8. Accounts of the Mahasi method may be found in Jack Kornfield, *Living Dharma: Teachings of Twelve Buddhist Masters* (Boston: Shambhala, 1995), 51–81; Mahasi Sayadaw, *Satipatthana Vipassana* (Seattle: Seattle Pariyatti Press, 1990); Nyanaponika Thera, *The Heart of Buddhist Meditation* (New York: Weiser, 1973) and *The Power of Mindfulness* (Unity Press); and E. H. Shattock's *Experiment in*

Mindfulness (New York: Weiser, 1972). The best account of Goenka's method is William Hart, *The Art of Living: Vipassana Meditation as Taught by S. N. Goenka* (San Francisco: HarperSanFrancisco, 1987).

9. Fronsdal, "Insight Meditation," in Prebish and Tanaka, eds., *Faces of Buddhism,* 173.

10. *Inquiring Mind* 18, no. 1 (fall 2001): 38.

11. At least two other figures associated with American Theravada/ vipassana should be mentioned here. Henepola Gunaratna is a Sri Lankan monk who took a Ph.D. in philosophy at American University and now leads the Bhavana Society, a practice center located in the Shenandoah Valley of West Virginia, based on Theravada monasticism but including American adaptations. Havanapola Ratanasara is also a university-educated Sri Lankan monk. After immigrating to the United States in 1980, he founded one of the earliest Theravada temples in Los Angeles, the Dharma Vijaya Buddhist Vihara. He currently holds many administrative and committee posts, including the presidency of the College of Buddhist Studies in Los Angeles. Grounded in Theravada monasticism, he is nevertheless working to move Buddhism in progressive directions, as evidenced by his efforts to revive full ordination for women. See Seager, *Buddhism in America,* chap. 9.

12. Batchelor, *Awakening,* 240.

AFTERWORD

1. The proper transliteration of Prof. Suzuki's first name is Daisetsu, but because the reading public has become accustomed to "Daisetz" we have opted for this usage.

2. Daisetz Suzuki, *Collected Writings on Shin Buddhism* (Kyoto: Shinshu Otani-Ha, 1973), 36.

3. This is a paraphrase of a quotation from the *Larger Sukhavati-vyuha Sutra (The Embellishment of the Pure Land)* as found in Kenryo Kanamatsu, *Naturalness: A Classic of Shin Buddhism* (Bloomington, Indiana: World Wisdom, Inc., 2002), 17–23. Kanamatsu's source, with slight adaptation, is F. Max Muller, *The Sacred Books of the East,* vol. XLIX (Oxford: Clarendon Press, 1894).

4. To come back to why I (Smith) volunteered to write this Afterword, the Nembutsu is deeply etched in my childhood memories. As I walked through the narrow lanes of the rural Chinese town in which I grew up, I would often hear passersby mumbling in our hillbilly Shanghai dialect, "na-ma-uh-mi-du-vah," which phonetically is remarkably close to the Japanese Nembutsu just cited in the text. In those boyhood years I did not know what the syllables meant, only that they had something to do with Buddhism, and it is gratifying in these later years to find myself returning full circle to those syllables and this time understanding them.

SUGGESTIONS FOR FURTHER READING: AN ANNOTATED GUIDE

∾∾∾

GENERAL ACCOUNTS

Although written in the 1920s, J. B. Pratt's *The Pilgrimage of Buddhism and a Buddhist Pilgrimage* (New York: AMS Press, 1928) remains a comprehensive, reliable, and readable account.

More recent and accessible is Richard Robinson and Willard Johnson's thorough and superbly informed *The Buddhist Religion*, 4th ed. (Belmont, CA: Wadsworth Publishing Co., 1997).

Peter Harvey's *An Introduction to Buddhism* (Cambridge: Cambridge University Press, 1990), gives the Robinson and Johnson volume a run for its money. Thorough, up-to-date, sympathetic, and readable.

Nancy Wilson Ross's *Buddhism: A Way of Life and Thought* (New York: Random House, 1981) remains as lovely and intelligent a general introduction to Buddhism as one might find, covering the Buddha's life, basic teachings, and three of Buddhism's major schools, Theravada, Tibetan Buddhism, and Zen. Modest length, beautiful plates, helpful maps, a glossary, and a bibliography add to its appeal, though the last of these is now somewhat dated.

Edward Conze's *Buddhism: Its Essence and Development* (1959; reprint, Birmingham, England: Windhorse Publications, 2002) still enlightens and charms in many ways. An oldie, but a goodie.

The section on Buddhism in Heinrich Zimmer's *The Philosophies of India*, edited by Joseph Campbell (New York: Bollingen Foundation, 1951), rivals Marco Pallis's *Peaks and Lamas* (see Vajrayana section of this bibliography) in the beauty of its diction and its uplifting power. This sentence in its opening pages sets the tone for the book: "The basic aim of any serious study of Oriental thought should be not merely the gathering and ordering of as much information as possible, but the reception of some serious influence."

The central thesis of Trevor Ling's fascinating and original *The Buddha: Buddhist Civilization in India and Ceylon* (New York: Scribner, 1973) is that the Buddha initiated not a religion, but a *civilization,* a plan for a total way of life embracing not only the private, inner world of individuals but the economic, social, and political practices of their society. Ling's sympathetic and erudite portrait of the Buddha relates him to the geographical, social, and economic environments of his time. Chapter 9 explains how the Buddha's social theory functioned during the rule of King Asoka and the notes to the chapter include a translation of Asoka's famous rock edicts.

Walpola Rahula's *What the Buddha Taught,* rev. ed. (New York: Grove, 1962), limits itself to a discussion of the Buddha's original teachings, but in this genre there is perhaps nothing finer or more concise. The main text is less than a hundred pages. Excellent selections from Buddhist scriptures enrich this revised edition.

An extraordinary blend of scholarship and dharma enthusiasm is *A Survey of Buddhism* (1957; reprint, Glasgow: Windhorse Publications, 1987), perhaps the most comprehensive of the over forty books written by one of the most prolific and influential Buddhists of our era, Dennis Lingwood, whose initiatic name is Sangharakshita.

For historical, anthropological accounts of Theravada Buddhism in Southeast Asia, we can think of no better place to start than Richard Gombrich's *Theravada Buddhism: A Social History from Ancient Benares to Modern Colombo* (London and New York: Routledge and Kegan Paul, 1988). Gombrich focuses on Sri Lankan Theravada. For rich accounts of Theravada in Thailand and Burma, see S. J. Tambiah's *Buddhism and the Spirit Cults in Northeast Thailand* (Cambridge: Cambridge University Press, 1970) and Melford Spiro's *Buddhism and Society: A Great Tradition and Its Burmese Vicissitudes* (New York: Harper & Row, 1970), respectively.

THE BUDDHA'S LIFE

Edwin Arnold's classic, *The Light of Asia* (Philadelphia: Altemus, 1879; reprint, Los Angeles: Theosophy Co., 1977), is a poetic rendition of the life of the Buddha that deservedly enjoyed its tremendous popularity in the late nineteenth and early twentieth centuries. One of the greatest stories ever told is here delivered with great power and feeling.

Bhikkhu Nanamoli's *The Life of the Buddha* (BPS Pariyatti Editions, first U.S. edition, 2001) is another classic compiled with great tenderness and insight by a Theravadin monk.

In David and Indrani Kalupahana's *The Way of Siddhartha: A Life of the Buddha* (Boston: Shambhala, 1982), a world-class Buddhist philosopher and his wife tell the Buddha's life with excellent selections from the Pali scriptures and a special emphasis on the Buddha's doctrine of dependent arising.

E. J. Thomas's *Life of the Buddha as Legend and History,* 3d rev. ed. (New York: Barnes and Noble, 1952), is a classic scholarly treatment.

R. A. Mitchell's *The Buddha: His Life Retold* (New York: Paragon House, 1991) succeeds in conveying the "inescapable imprint of the Buddha's warm and forceful personality [and the] time-defying power of his unsurpassed intellect and unique spiritual consciousness."

THE PALI CANON AND THE DHAMMAPADA

The Buddha's discourses are grouped in five *nikayas,* or collections: the *Samyutta,* the *Majjhima,* the *Digha,* the *Anguttara,* and the *Khuddhaka.*

Bhikkhu Bodhi's *The Connected Discourses of the Buddha* is a fresh, authentic translation of the *Samyutta Nikaya* (London: Wisdom Publications, 2002), 2,080 pages; Bhikkhu Nanamoli and Bhikkhu Bodhi's *Middle Length Discourses of the Buddha* is a new translation of the *Majjhima Nikaya* (London: Wisdom Publications, 1995), 1,424 pages; and Maurice Walshe's *The Long Discourses of the Buddha* is a new translation of the *Digha Nikaya* (London: Wisdom Publications, 1987), 656 pages.

The richest small anthology of Pali scriptures that we know is Nyanatiloka's *The Word of the Buddha* (Kandy, Sri Lanka: Buddhist Publication Society, 1981).

Some feel that the nineteenth-century translations in Henry Warren's *Buddhism in Translation* (reprint, Cambridge, MA: Harvard University Press, 1953) have never been surpassed, while others find their elegance compromised by age.

The *Dhammapada* ("Path of Dhamma"), a collection of some four hundred sayings attributed to the Buddha, is the world's best-known part of the Pali Canon's *Khuddaka Nikaya.* Our favorite translation is Eknath Easwaren's *The Dhammapada* (Tomales Bay, CA: Nilgiri Press, 1985, 1996). His Indian heritage, literary gifts, and spiritual sensibilities (which have given us excellent translations of Hinduism's *Upanishads* and *Bhagavad Gita*) here produce a sublime rendering of the words of the Buddha. Verse after verse shimmers with quiet, confident authority. A bonus is the sparkling 70-page introduction to the Buddha's life and teachings that precedes the translation. Also highly recommendable are Irving Babbitt's *The Dhammapada* (New York: New Directions, 1965) and Narada Thera's *The Dhammapada* (Colombo: Vajirarama, 1972).

MAHAYANA SCRIPTURES

Edward Conze's slim volume *Buddhist Wisdom Books* (London: Allen and Unwin, 1958) is a discerning translation of and commentary on two of the most famous Mahayana scriptures, the Diamond Sutra and the Heart Sutra.

A excerpt from the Sukhavati Sutra, the scriptural basis of the Pure Land tradition, may be found in Edward Conze's *Buddhist Scriptures* (Baltimore: Penguin Books, 1959).

Marion Matics's *Entering the Path of Enlightenment* (New York: Macmillan, 1970) is a lovely translation of probably the single most cherished expression of the bodhisattva ideal, Santideva's eighth-century poem the *Bohdicaryavatara*. Matics's 140-page introduction is rich and helpful.

Sangharakshita's *The Drama of Cosmic Enlightenment: Parables, Myths and Symbols of the White Lotus Sutra* (Glasgow: Windhorse Publications, 1993) is a series of informative lectures on the Lotus Sutra, an important Mahayana text.

PAN-BUDDHIST SCRIPTURES

Any one of the following texts will give interested readers a broad, reliable sense of the range of Buddhist sacred literature: Edward Conze, *Buddhist Scriptures* (Baltimore: Penguin Books, 1959); Edward Conze et al., *Buddhist Texts Through the Ages* (Oxford, England: Oneworld Publications, 1995); John Strong, *The Experience of Buddhism: Sources and Interpretations* (Belmont, CA: Dickenson, 1973); William T. De Bary, ed., *The Buddhist Tradition in India, China, and Japan* (New York: Modern Library, 1969); E. A. Burtt, ed., *The Teachings of the Compassionate Buddha* (New York: New American Library, 1955); and Sangharakshita, *The Eternal Legacy* (London: Tharpa, 1985).

A new anthology of Buddhist writings has been created by Donald J. Lopez, *A Modern Buddhist Bible: Essential Writings from East and West* (Boston: Beacon Press, 2002).

BUDDHIST PHILOSOPHY

T. R. V. Murti's *The Central Philosophy of Buddhism* (New York: Macmillan, 1955) is neither easy nor uncontroversial, but for Buddhists who like to think, it is surely one of the best of the twentieth century's books on Buddhism, a tenacious exposition of the Madhyamika ("Middle Way") School and its doctrine of emptiness.

K. V. Ramanan's *Nagarjuna's Philosophy* (1966; reprint, Columbia, Missouri: South Asia Books, 1993) is a rich and wonderful study of the philosopher who has been called "the second Buddha."

David Kalupahana's *Nagarjuna: The Philosophy of the Middle Way* (Albany: State University of New York Press, 1986) is a new translation of Nagarjuna's major work prefaced by a vigorous 100-page argument that reassesses Nagarjuna's place in the history of Buddhist philosophy.

David Kalupahana's *A History of Buddhist Philosophy: Continuities and Discontinuities* (Honolulu: University of Hawaii Press, 1992) is the mature work of one the greatest historians of Buddhist philosophy in our era.

David Loy's *Lack and Transcendence: The Problem of Death and Life in Psychotherapy, Existentialism and Buddhism* (Atlantic Highlands, NJ: Humanities Press, 1996) is a magnificent set of essays by a Zen-trained philosopher. Loy argues that the basic repression in human psychological life is not sex (à la Freud) or death (à la the neo-Freudian Ernest Becker), but the fact of anatta, or no-self. Armed with this diagnosis, he thoughtfully critiques both the Western philosophical tradition and contemporary culture.

Nyanaponika Thera's *The Vision of Dhamma* (York Beach, ME: Samuel Weiser, 1986) is for the connoisseur. It is a series of profound, pellucid essays on key Buddhist ideas by one of the great Theravadin scholar-monks of the twentieth century.

Edward Conze's *Buddhist Thought in India* (London: Allen and Unwin, 1962), though definitely not for beginners, is a penetrating study of the evolution of early Buddhist ideas by this celebrated twentieth-century scholar.

Though Anagarika Govinda spent most of his life in the Vajrayana, an early work, *The Psychological Attitude of Early Buddhist Philosophy* (London: Rider, 1969), remains an excellent, engaging exposition of the early Buddhist psychology, replete with interesting illustrations.

Nolan Pliny Jacobson's *Understanding Buddhism* (Carbondale: Southern Illinois University Press, 1986) is one of the most beautiful and least-known books on Buddhism in English, a series of fresh, often stunningly original, essays.

Stephen Batchelor's *Buddhism Without Beliefs* (New York: River-head, 1997) is a series of spare, intelligent essays on Western Buddhism that raises the issue of how much of traditional Buddhism is dispensable baggage. Batchelor is one of the brightest, most well informed, and widely experienced of contemporary Buddhist authors and a reliable guide.

Garma C. C. Chang's *The Buddhist Teaching of Totality* (University Park: Pennsylvania State University Press, 1974) remains an accessible and delightful examination of the Buddhist idea of interdependence.

PURE LAND BUDDHISM

A book that takes readers by the hand, so to speak, and leads them straight to the heart of Pure Land Buddhism is Hiroyuki Itsuki's *Tariki* (Tokyo and New York: Kodansha, 2001). Its multiple virtues center in the fact that it is a candid, autobiographical account of someone who had been brought to the brink of despair by a series of cataclysmic misfortunes and is redeemed by the Other Power *(tariki)* that comes to him through Pure Land Buddhism. That the author is one of Japan's foremost novelists ensures that it is beautifully written.

At the other extreme stands *The Pure Land Tradition: History and Development*, ed. James Foard, Michael Solomon, and Richard Payne (Berkeley Buddhist Series, Berkeley, CA: Regents of the University of California, 1996). A veritable warehouse of factual information, it recounts more about Pure Land than general readers will feel they need to know, but is the best single reference book on the subject. Readers who would like to venture one step beyond what our book says about Pure Land in its Afterword will be served by Taitetsu Unno's chapter titled "Shinran: A New Path to Buddhahood."

Between the above two books stands Kenro Kanamatsu's *Naturalness: A Classic of Shin Buddhism* (World Wisdom, P.O. Box 2862, Bloomington, IN, 47402). It is a lovely, slim volume that can be read at a sitting and that conveys the essence of Pure Land beautifully.

BUDDHIST MEDITATION

"Better than a thousand useless words is one single word that gives peace," said the Buddha. Buddhist meditation teachers similarly seem to suggest that "better than a thousand books on meditation is one single meditation course." There are now many places in the West where excellent meditation training is offered. See Chapters 16, 17, and 18 for help in this regard. The most comprehensive listing of Buddhist centers and groups is Don Morreale's *Complete Guide to Buddhist America* (Boston and London: Shambhala, 1998).

Vipassana

An early, classic account of *satipatthana* ("mindfulness") meditation (a.k.a. vipassana) is Nyanaponika Thera's *The Heart of Buddhist Meditation* (New York: Weiser, 1973).

William Hart's *The Art of Living: Vipassana Meditation as Taught by S. N. Goenka* (San Francisco: HarperSanFrancisco, 1987) is an excellent introduction to this form of practice. See Chapter 18.

The influential vipassana meditation methods of the Burmese master Mahasi Sayadaw (see Chapter 18) are described in Jack Kornfield's *Living Dharma: Teachings of Twelve Buddhist Masters* (Boston: Shambhala, 1995), 51–81; Mahasi Sayadaw's *Satipatthana Vipassana* (Seattle: Pariyatti Press, 1990); Nyanaponika Thera's above mentioned *The Heart of Buddhist Meditation* and *The Power of Mindfulness* (Seattle: Pariyatti, 1986); and E. H. Shattock's *Experiment in Mindfulness* (New York: Weiser, 1972).

Fine introductions to the Buddhist practices of the Insight Meditation Society and the Spirit Rock Meditation Center (see Chapter 18) can be found in Joseph Goldstein's *Insight Meditation: The Practice of Freedom* (Boston: Shambhala, 1994); Goldstein, Kornfield, et al.'s *Seeking the Heart of Wisdom: The Path of Insight Meditation* (Boston: Shambhala, 2001); Sharon Salzberg's *A Heart as Wide as the World* (Boston: Shambhala, 1997); Sylvia Boorstein's *Don't Just Do Something, Sit There: A Mindfulness Retreat* (San Francisco: HarperSanFrancisco, 1996); and Henepola Gunaratana's *Mindfulness in Plain English* (London: Wisdom Publications, 1993).

Two superb films on the practice of vipassana in prisons that also serve as excellent introductions to this form of meditation are *Doing Time, Doing Vipassana,* and *Changing from Inside,* both available from pariyatti.com.

Zen and Zazen

Two very different American Zen classics that complement each other beautifully are Philip Kapleau's *The Three Pillars of Zen* (New York: Anchor Books, 1989) and Shunryu Suzuki's *Zen Mind, Beginner's Mind* (New York: John Weatherhill, 1970).

Irmgard Schloegel's *Wisdom of the Zen Masters* (New York: New Directions, 1975) is an instructive and delightful collection of Zen sayings.

There will probably never be a book written in English that takes one inside the koan training of Rinzai Zen as far as does Isshu Miura and Ruth Fuller Sasaki's *Zen Dust* (New York: Harcourt, Brace & World, 1966).

William Barret's *Zen Buddhism: Selected Writings of D. T. Suzuki* (Garden City, NY: Doubleday, 1956) is a fine selection of essays by the Japanese scholar who made Zen an English word.

Robert Aitken's *The Dragon Who Never Sleeps: Verses for Zen Buddhist Practice* (Berkeley, CA: Parallax, 1992), *Taking the Path of Zen* (San Francisco: North Point Press, 1982), and *Mind of Clover: Essays on Zen Buddhist Ethics* (San Francisco: North Point Press, 1984) are among the many thoughtful works of this first American Zen master.

Carl Bielefeldt's *Dogen's Manuals of Zen Meditation* (Berkeley: University of California Press, 1988) is a learned point of entry into Dogen studies by a Stanford University professor.

John Blofeld's *The Zen Teaching of Huang Po* (New York: Grove Press, 1958) and [Hui Hai's] *Zen Teachings of Instantaneous Awakening* (Buddhist Publication Group, 1994) are vivid translations by an Englishman who spent the greater part of his life studying Chinese Buddhism.

Thich Nhat Hanh's *Miracle of Mindfulness: A Manual on Meditation* (Boston: Beacon Press, 1992) is a fine example of the gentle, profound teaching of this internationally known peace activist and Buddhist monk.

Hubert Benoit's *Zen and the Psychology of Transformation* (formerly *The Supreme Doctrine*), (Rochester, VT: Inner Traditions International, 1990) is definitely not for everyone. Despite the seemingly fatal error of approaching Zen through categories drawn from Gurdjieff and Vedanta, this French psychiatrist produces a stunningly insightful, utterly original book. Heady and rich.

VAJRAYANA AND TIBETAN BUDDHISM

Senior Vajrayana scholar Reginald Ray's *Indestructible Truth: The Living Spirituality of Tibetan Buddhism* (Boston and London: Shambhala, 2000) and *Secrets of the Vajra World* (Boston and London: Shambhala, 2001) have appeared too recently for us to carefully review, but are winning high praise from informed readers.

Robert Thurman's *Inner Revolution: Life, Liberty and the Pursuit of Real Happiness* (New York: Riverhead Books, 1998) is this prominent and ebullient scholar's envisagement of a "second" American Revolution through which Buddhist ways of life will complete the American experiment in democracy and fulfill, albeit in unexpected ways, the American dream of genuine freedom and happiness. Among Thurman's other works is his recent, excellently introduced translation of the most famous of all Tibetan texts, *The Tibetan Book of the Dead:*

Liberation Through Understanding in the Between (New York: Bantam, 1994).

Tenzin Gyatso, the fourteenth (and current) Dalai Lama, is the author of numerous books, many of them published by Snow Lion Press and Wisdom Publications. A few examples: *The Buddhism of Tibet and the Key to the Middle Way,* translated by Jeffrey Hopkins (London: Allen and Unwin, 1975); *Kindness, Clarity and Insight,* translated by Jeffrey Hopkins (Ithaca, NY: Snow Lion, 1984); and *Freedom in Exile* (New York: Harper Perennial, 1990).

The late, influential Tibetan teacher Chogyam Trungpa's *Cutting Through Spiritual Materialism* (Berkeley, CA: Shambhala, 1973*)* is a contemporary classic and perhaps this author's most enduring work. He also cotranslated a version of *The Tibetan Book of the Dead.*

Lama Anagarika Govinda's *Foundations of Tibetan Mysticism* (York Beach, ME: Samuel Weiser, 1969) presents the theory of Tibetan Buddhism, while Marco Pallis's *Peaks and Lamas* (London: Woburn Press, 1974) is one of the finest spiritual travelogues ever written.

Huston Smith's half-hour videotape on Tibetan Buddhism, "Requiem for a Faith," makes available the audiovisual dimensions of the Vajrayana as described in Chapter 10 of this book. It can be secured from The Hartley Film Foundation, Cat Rock Road, Cos Cob, CT 06807.

Sogyal Rinpoche's *The Tibetan Book of Living and Dying* (San Francisco: HarperSanFrancisco, 1992) is an exposition of Vajrayana thought and practice that has enjoyed a wide audience (Revised edition, 2002).

Walt Anderson's *Open Secrets: A Western Guide to Tibetan Buddhism* (New York: Penguin, 1979) remains a helpful, highly readable account. An excellent place to begin.

At 450 pages, John Powers's *Introduction to Tibetan Buddhism* (Ithaca, NY: Snow Lion, 1995) is a thorough, serviceable general account.

Donald S. Lopez's *Prisoners of Shangri-La: Tibetan Buddhism in the West* (Chicago: University of Chicago Press, 1998) is an erudite critical reflection on the Western danger of romanticizing this tradition.

BUDDHISM IN THE WEST AND IN AMERICA

Rick Fields's *How the Swans Came to the Lake: A Narrative History of Buddhism in America,* 3d ed. (Boston and London: Shambhala, 1992) and Stephen Batchelor's *The Awakening of the West: The Encounter of Buddhism and Western Culture* (Berkeley, CA: Parallax Press, 1994) are excellent historical narratives studded with jewels of

Buddhist insight. Fields covers the American ground, Batchelor the European.

The Faces of Buddhism in America, edited by C. Prebish and K. Tanaka (Berkeley: University of California Press, 1998), is a collection of sixteen essays on the various forms of American Buddhism and the social and philosophical issues they face.

James Coleman's *The New Buddhism: The Western Transformation of an Ancient Tradition* (New York: Oxford University Press, 2001) is a most helpful and sympathetic sociological study of American Buddhism.

Charles Prebish's *Luminous Passage: The Practice and Study of Buddhism in America* (Berkeley: University of California Press, 1999) is the excellent and most recent work of an experienced chronicler of American Buddhism.

Richard Hughes Seager's *Buddhism in America* (New York: Columbia University Press, 1999).

Al Rapaport's *Buddhism in America: Proceedings of the 1997 Conference on the Future of Buddhist Meditative Practices in the West* (Rutland, VT: Tuttle, 1998).

Don Morreale's *The Complete Guide to Buddhist America* (Boston and London: Shambhala, 1998).

Martin Baumann's "The Dharma Has Come West: A Survey of Recent Studies and Sources," *Journal of Buddhist Ethics* 4 (1997).

See also Helen Tworkov's *Zen in America: Five Teachers and the Search for an American Buddhism* (New York: Kodansha International, 1994) for profiles of Robert Aitken, Jakusho Kwong, Bernard Glassman, Maurine Stuart, and Richard Baker.

WOMEN IN BUDDHISM

Three books that reveal the important role women are playing in the unfolding of American Buddhism are Sandy Boucher's *Turning the Wheel: American Women Creating the New Buddhism* (San Francisco: HarperSanFrancisco: 1988), Lenore Friedman's *Meetings with Remarkable Women: Buddhist Teachers in America* (London and Boston: Shambhala, 1987), and Karma Lekshe Tsomo's *Buddhism Through American Women's Eyes* (Ithaca, NY: Snow Lion, 1995).

Other books on various aspects of this topic include: Kathryn Blackstone's *Women in the Footsteps of the Buddha: Struggle for Liberation in the Therigatha* (Surrey, England: Curzon, 1998); Tessa J. Bartholomeusz's *Women Under the Bo Tree: Buddhist Nuns in Sri Lanka* (New York: Cambridge University Press, 1994); and Sandy Boucher's *Opening the Lotus: A Women's Guide to Buddhism* (Boston:

Beacon, 1997); Marianne Dresser's *Buddhist Women on the Edge: Perspectives from the Western Frontier* (Berkeley, CA: North Atlantic Books, 1996); Rita Gross's *Buddhism After Patriarchy: A Feminist History, Analysis and Reconstruction of Buddhism* (Albany: State University of New York Press, 1993); Anne Klein's *Meeting the Great Bliss Queen: Buddhists, Feminists and the Art of the Self* (Boston: Beacon, 1996); Susan Murcott's *The First Buddhist Women: Translations and Commentaries on the Therigatha* (Berkeley, CA: Parallax, 1991); Diana Paul's *Women in Buddhism: Images of the Feminine in Mahayana Buddhism* (Berkeley: University of California Press, 1985); Miranda Shaw's *Passionate Enlightenment: Women in Tantric Buddhism* (Princeton, NJ: Princeton University Press, 1995); and Karma Lekshe Tsomo's *Sakyaditta: Daughters of the Buddha* (Ithaca, NY: Snow Lion, 1988).

WEB SITES

A magisterial compilation and discussion of the Cybersangha, i.e., Buddhist Web sites, can be found in Chapter 4 of Charles Prebish's *Luminous Passage: The Practice and Study of Buddhism in America* (Berkeley: University of California Press, 1999). As most Web sites contain links to numerous others, we name only a few that we have found particularly helpful:

http://www.baumann-martin.de-, the homepage of Martin Baumann, a European Buddhist scholar who chronicles the global Buddhist scene

http://www.jbe.gold.ac.uk (jbe = *The Journal of Buddhist Ethics*)

http://www.accesstoinsight.org

http://www.dharmanet.org (Dharma Net International)

http://www.buddhanet.net

INDEX

~~~

Abhidhamma, 75, 205*n*. 2
Achaan Sumehdo (Robert Jackman), 131–32
Aitken Roshi, Robert, 151, 157–58, 215*n*. 10
Amaro, Ajahn, 181–82
America, Buddhism in, 136–42, 212*n*. 1; Abhayaghiri Buddhist Monastery, CA, 182; American Institute of Buddhist Studies, 163; Bhavana Society, WV, 218*n*. 11; Buddhist Churches of America, 141; Chinese, 140–41; Community for Mindful Living, 160; Desert Vipassana Center, 174; Dharma Realm Buddhist University, 141; Dharma Vijaya Buddhist Vihara, Los Angeles, 218*n*. 11; Diamond Sangha, 158; ethnic Buddhism, 140–41, 213–14*n*. 9; evangelical Buddhism, 141–42; Ewam Choden Tibetan Buddhist Center, Kensington, CA, 165; first American to join a Buddhist Sangha on American soil, 139; First Zen Institute of America, 150–51, 158; God Mountain Monastery, 141, 214*n*. 10; Hsi Lai Temple, 140; Insight Meditation, 174, 217*n*. 2; Insight Meditation Society (IMS) and Mahasi approach, 175–76, 178–81, 217*n*. 4, 217*n*. 5, 217*n*. 8; Japanese in Hawaii, 141; Korean immigrants, 141; Korean Zen, 159–60; Koto-an Zendo, Hawaii, 158; Lama title and retreat, 169, 216–17*n*. 8; Mahayana Buddhism, 141; Maui Zendo, 158; meditation, vipassana, 172–83; meditation, Zen, 154–60; Namgyal monastery, Ithaca, NY, 164; Naropa Institute, Boulder, CO, 167, 175; New Buddhism," 140, 143–49, 213*n*. 9; Nichiren Shoshu of America, 142; number of Buddhists in, 139–40; Nyingma Meditation Center, Berkeley, CA, 165–66; Osel Shen Phen Ling, Missoula, 164; Pure Land Buddhism, 141, 213*n*. 9; Rigpa Fellowship, Santa Cruz, CA, 166; Sakya Monastery of Tibetan Buddhism, 165; Sakya Phunstok Ling, Silver Springs, MD, 165; Sakya Sheidrup Ling, Cambridge, MA, 165; Shambala International, 167; Shasta Abbey, 159; Sino-American Buddhist Assocation, 140–41; Soka Gakkai, 141–42; Soto Zen Mission, San Francisco, 154; Spirit Rock Meditation Center, Woodacre, CA, 175–76; Springwater Center, NY, 159; Theosophical Society, 138; Theravada Buddhism, 141; Thoreau and Buddhism, 136–37; Tibet House, 163; Tibetan Buddhism in exile, 163–71; Tibetan Buddhist Learning Center, 161–62, 164, 216*n*. 2; University of Oriental Studies, 160; University of Wisconsin, Buddhist Studies, 162, 164; Vietnamese immigrants, 141; vipassana centers (Goenka-ji), 177–78); vipassana movement, 172–83, 217–18*n*. 8; World Parliament of Religions, 1893, 138–39,